Table of Contents

KETO COACH

HANDBOOK

Including Simplified Science

And

Recipes

By Sergio Guzzardi

HYBRID Publishers

Australia

COPYRIGHT

Second Edition, July 2019

ISBN: 9781983047886

HYBRID Publishers

PO BOX 52, Ormond,

Victoria, 3204

Australia

Introduction

———

————————

THERE ARE A LOT OF misconceptions about aging, and the first and foremost is that we're doing better than ever in terms of longevity. Yes, the average life expectancy has increased over the last five decades. In 1960, the average life expectancy of western men was 66.4 years; by 2013, it was a full ten years longer. For women, the average ages were 73.1 and 81.1, respectively. But much of this extended life span can be attributed to the fact that we've developed vaccines, antibiotics, and hygiene protocols to beat the main causes of a shortened average life span, namely infectious diseases that disproportionately affected young children. Perhaps we've come to an end in terms of what modern advances can accomplish. Sadly, life expectancy has now declined for the last

three years! And never forget, people have lived into very old age since recorded time. Today we're starting to see both a decreasing life span and a horribly reduced health span, the length of time people maintain full function. Most people now see their health begin to decline at age 50. Yet we've gotten very good at extending our life span with a host of medical procedures, drugs, and treatments. So we're living longer, but we're not living better

PART ONE: THE SCIENCE BEHIND THE KETOGENIC DIET

———

Chapter 1

Basics of the Ketogenic Diet

OF THE MACRONUTRIENTS, carbohydrates are therefore argued to be the major cause of weight gain. This is more so because the increased intake of high glycemic index carbohydrate foods generally causes fluctuating blood sugar levels due to their fast absorption into the bloodstream and which more often than not leads to the overproduction of insulin. This is where the problem actually starts.

Insulin is a hormone that regulates blood glucose levels and therefore maintenance of the energy in/energy out equation of the body which rules body weight. Excess amounts of glucose in the bloodstream causes the excessive secretion of insulin which leads to the storage of the excess glucose in the body as either glycogen in liver and muscle cells or fat in fat cells.

Despite the ability of ketogenic diets to reduce insulin production, their main objective is ultimately aimed at inducing the state of ketosis. Ketosis can be regarded as a condition or state in which the rate of formation of ketones produced by the break down of "fat" into "fatty acids" by the liver is greater than the ability of tissues to oxidize them. Ketosis is actually a secondary state of the process of lipolysis (fat break down) and is a general side effect of low-carbohydrate diets. Ketogenic diets are therefore favorably disposed to the encouragement and promotion of ketosis.

Prolonged periods of starvation can easily induce ketosis but it can also be deliberately induced by making use of a low-calorie or low-carbohydrate diet through the ingestion of large amounts of either fats or proteins and drastically reduced carbohydrates. Therefore, high-fat and high-protein diets are the weight loss diets used to deliberately induce ketosis.

Essentially, ketosis is a very efficient form of energy production which does not involve the production of insulin as the body rather burns its fat deposits for energy. Consequently, the idea of reducing carbohydrate consumption does not only reduce insulin production but also practically forces the body to burn its fat deposit for energy, thereby making the use of ketogenic diets a very powerful way to achieve rapid weight loss.

Ketogenic diets are designed in such a way that they initially force the body to exhaust its glucose supply and then finally switch to burning its fat deposits for energy. Subsequent food intakes after inducing the state of ketosis are meant to keep the ketosis process running by appropriately adjusting further carbohydrate consumption to provide just the basic amount of calories needed by the body.

Your body needs energy to run properly, and it gets this energy from the food you eat. Food is turned a form of sugar called glucose. Your cells (and particularly the ones in your brain) need a constant supply of glucose, and if it gets low you begin to feel fatigued and weak.

Glucose circulates in your blood after you eat, and it is used up fairly rapidly as you go about your everyday tasks. If not replenished, it is, in fact, depleted in a few hours. This creates a problem: how do you maintain a good supply? Glucose itself can't be stored, but it can be turned into a form called glycogen that can be stored in your muscles and liver. From here it can be drawn out and used as needed. It is usually good for about 10 to 12 hours.

What happens when it is depleted? The body then turns to the fat cells that are stored throughout your body. They can be broken down and converted to what is called ketones. This is, of course, what dieters look for, namely, the loss of fat cells. But you have to be careful if you remain in this stage for too long. The body soon begins to break down protein; it can also be converted to glucose through a rather complicated process. And this causes the loss of muscle - something you don't want. Indeed, in most diets, a fair amount of the weight loss comes from muscle loss along with depletion of water (leaving you dehydrated). So don't be deceived.

Starting a ketogenic diet

Before starting on a low carb eating plan, educate yourself on the types of foods you can eat. It is also helpful to have a quick reference guide for counting the grams of carbs in various foods. In the beginning stages of the diet, you will want to limit your daily net carb count at about 30 grams.

Net carbs are the total carbs minus the fiber. The daily carb intake varies so you will need to determine the best level for you to continue losing weight. The key is to eat fewer carbohydrates than your body burns in a day. Typically, the carbs can be slowly increased as long as weight loss continues.

People with slower metabolisms will need to keep the carb count extremely low for the duration of the diet in order to continue to lose weight. Taking a supplement to boost metabolism or consulting a doctor regarding prescription medications may be required if ongoing weight loss becomes difficult.

Preparing your own meals is less costly than buying prepared foods and it is easier to keep track of the carb counts. The internet is full of free low carb recipes and there are a number of good low carb cookbooks. Learn the basics of low carb cooking and the types of ingredients used.

Almond meal and coconut flour are excellent substitutes for flour. Both natural and artificial sweeteners are available to replace sugar in dessert recipes. Having a good collection of recipes on hand will get you off to a good start when planning meals.

Now is a good time to go through the food in your kitchen to get rid of anything that may be a temptation. Items that contain flour and/or sugar should be eliminated. If other family members insist on having high carb foods around, you will need to come up with a plan to stay away from them.

The good news is that there are delicious low carb substitutes to replace these forbidden foods. Having plenty of great tasting low carb foods around will prevent the urge to eat sugary snacks.

You will need to read food labels while shopping. A good rule of thumb is to stick with foods that are less than 10 net grams of carbs per serving. While in the grocery store, your shopping should be focused on the perimeter of the store. The aisles are loaded with processed foods that are usually high in carbs. The primary low carb staples are protein rich foods such as eggs, cheese and meat.

It's also important to eat plenty of low carb vegetables. Use a carb counting reference guide to determine which foods to choose. Transitioning to a lower carb diet may seem difficult at first, but remember things will get easier. There is no need to worry about the amount of food as long as the net carbs are within reason.

Be sure to drink plenty of water as it is helpful to keep the body flushing out wastes. Weigh yourself every few days to make sure you are continuing to lose weight. Adjust your daily net carb intake as needed to stay on track.

The Keto diet involves going long spells on extremely low (no higher than 20g per day) to almost zero gram per day of carbs and increasing your fats to a really high level (to the point where they may make up as much as 75% of your daily macronutrients intake.)

Depleting your carbohydrate/glycogen liver stores and then moving onto fat for fuel means you should end up being shredded.

_____

GRAIN CONTROVERSY

Most of our Western Governments offer health guidelines which ask us to base our food intake almost universally around grain-type carbohydrates, what were once grouped as starches. We know these most commonly as rice, pasta, potatoes and breads.

These types of food appear to have been staples of our western diets since time immemorial (they're not, but that's another story). We are often told that eating these foods will leave us full, satisfied and full of a slow releasing stream of energy that is healthy and safe. Unfortunately, at least for human beings, this doesn't always appear to be the case.

Not all grains are created equal for a start and this can be where grain advocates purposely or accidentally mislead. For instance most rice, particularly white rice, will convert to sugar almost immediately in our system and we've already seen some of the devastating effects of excess sugar consumption. Grains, no matter what source they come from will cause elevated insulin levels.

For the very healthy amongst us, who have extremely sensitive insulin (either through good genetics, regular exercise or a combination of both) may be able to carefully use small quantities of grains to fuel their bodies through the periods of high activity. However for the vast majority of people, the excess of grains will result in almost all the same problems as sugar consumption.

Many low-carb exponents are suspicious of medical advice to eat grains, many citing government subsidies of mass agriculture. Eating grains is a very cheap and simple way of providing food, but cheap and simple is rarely the same as healthy and good.

Okay, So Are There Any Good Carbohydrates?

The good news for carbohydrates is that they can claim the healthiest of foodstuffs amongst their number. Here is where a simple view of the Atkins diet and other strict no or low carb diets flounder. Not all carbs are created equal as we know and the carb group that is utterly essential to our survival?

Vegetables

Low carb diets have often been seen as lacking in vegetables as people carefully trim away all excess carbohydrates, effectively throwing the baby out with the dirty bathwater. On the subject of vegetables you won't find much dissension amongst medical experts of any standpoint.

These wonderful foodstuffs not only contain a plethora of vitamins and minerals, but also are often chock-full of fibre, water and a host of exotic cancer-fighting substances unique to vegetables.

The important thing about vegetables is that they are nutrient dense and calorie sparse. In plain English, they contain a lot of good stuff in a very small package. You can eat virtually enough vegetables to fill you up and still have eaten only a tiny percentage of the calories a normal diet would confer.

One of the arguments for regular grain consumption is the necessary vitamins and minerals they contain, not to mention the essential fibre for our digestive tract. But guess what? Vegetables make grains seem pretty redundant.

A small handful of organic vegetables will contain more vitamins and minerals than virtually a day's worth of grains, all in an easier to digest package, with extra water and no danger of insulin overload.

Even on a low-carb diet you can stuff yourself silly with vegetables without fear. The primary advantage of a low-carb diet is insulin control and vegetables won't interfere with that.

Remember organic vegetables have a much higher vitamin and mineral content, also the darker green or red a vegetable the higher the amount of beneficial Chlorophyll inside the plant. Try to eat your veggies raw and fresh and often. A regular supply of varied veggies is like nature's most perfect multivitamin pill.

To be a healthy low carber you need to investigate healthy fats a little more and remember that high quality, preferably organic oils are a better choice than others. There are a host of books on this subject and a host of great products out there.

Unfortunately due to the mass pollution of the seas, fish may no longer be the healthiest option, although carefully filtrated fish-oils (by Companies who are clued up on the science of keeping these oils in a health-giving state) are widely available and a must-buy for everyone.

Protein covers the widest range of foods left to us. Protein, which makes up our body's muscles, can be found from the flesh of other animals as well as from milks, beans and lentils. Much like fat, our body requires protein. How much is open to debate.

Active individuals, particularly those who require larger muscles, will have a much higher protein need than a sedentary individual but sufficed to say, excess protein intake (although feared by many mainstream nutritionists) has none of the dangers that excess grain or sugar consumption does.

That said, we could always make healthier choices. Although the Atkins diet may allow us to eat burgers and bacon all day long, this may not be the ideal choice. When considering meat products we have to remember what state the animal it came from was in when it was slaughtered.

Most animals in large factory farming business are over-fed, over medicated cripples and surely this meat can't be entirely healthy. Foods like bacon also contain a large number of hazardous preservative chemicals that sap at our besieged immune systems. Once again, not all proteins are created equal. Choosing organic fresh white meats is a wise choice when considering health.

Chicken and Pork, from good organic sources is a lean and easy to use protein source. Animals such as buffalo and ostrich may sound like exotic food sources to many, but their meat is almost entirely free from chemicals and their natural diets of grass and other non-artificial feeds leaves them with a low-fat content of good, healthy fats.

High quality protein is essential to your health and survival. Eating lower-quality meats may allow you to stay trim (since protein consumption appears to regulate our appetite much better than grains ever could) but investing in higher quality meats will mean you can claim the health benefits as well.

The healthy low carb approach

As many low-carb dieters have pointed out, most humans were never designed to live on a high carbohydrate content in their diets. As hunter-gatherers we consisted mostly on animals that roamed wild and on fresh vegetables and berries we could find in our local habitat.

Although our societies may have advanced enough to let us devise sustained agriculture, our genes are still locked in a hundred thousand-year-old struggle for survival. Our bodies recognize the nutrients available from clean meats, healthy fats and fresh vegetables. They have substantial trouble coping with the sudden influx of excess energy and too quickly absorbed carbohydrates in the form of grains and sugars.

Restricting the intake of grains and sugars makes a fairly quick and positive change towards a healthier life. However, it may be that in our urge to shed the pounds with as little pain as possible, the lower carb diets we choose are tilted towards the proteins and fats we don't really need and, attention to vegetables is ignored.

With a few minor modifications we can find a lower-carbohydrate approach that not only helps us maintain a normalized body-weight and fat mass but also helps us be an all-round healthier individual. There are a hundred other points towards improving health but all these changes make an admirable start.

The basis of many diets in the past was, "If you want to lose fat, you have to lower the fat intake to minimum." That was also what the leading health institutes recommended. But in the last few years that thesis was proven wrong.

There are many scientific studies that show that we lose fat faster if we limit our carbohydrate consumption to minimum, while increasing the fat intake.

Sources of energy for the body

Body needs energy for its activity. In general body gets this energy primary from carbohydrates, then fat and if necessary also from protein. So if you limit the carbohydrate intake to 50grams or less, your body will have to search for alternative fuel source, which is fat. Your body can still manufacture carbohydrates from protein and one of the components of fat (glycerol).

Every organ except the brains and the nervous system can use fatty acids as an alternative fuel source. Actually the brains and the nervous system can function fairly efficiently without glucose (carbohydrates), because they can get up to 75% of energy from ketones. Ketones are a byproduct of the incomplete breakdown of fatty acids in the liver. They are used as fuel for brains and nervous system.

If you stay on the ketogenic diet for few days, the body starts manufacturing more and more ketones and it greatly reduces the usage of glucose.

At the same time the conversion of protein to energy is reduced, which is very important when trying to keep lean body mass (muscle). More muscle mass you have, the more calories your body will burn. This is one of the reasons ketogenic diet is so effective.

Body needs three weeks to completely adjust to the usage of fatty acids and ketones as an energy source. If you try the ketogenic diet the first few days will be like hell, you will not be able to concentrate, you will be nauseous and weak.

This is because the body needs some time to adjust to the new energy source. The body is used to carbohydrates and if you just lower your carbohydrate to less then 20g, that will be a big shock. But when the body adjusts you will start benefiting from the ketogenic diet.

When you start eating high fat and low carbohydrate diet you will start influencing mainly two important hormones, insulin and glucagon. Insulin in the body transports nutrients from blood to cells (like glucose to muscles).

Glucagon acts as an opposite, it influences the cells to starts releasing the stored nutrients to the blood stream. When there is a shortage of glucose it encourages the lives to produce glucose from other sources and releasing them to the blood, where they can reach any cell in the body.

If you lower the carbohydrate intake, the body gradually starts releasing less and less insulin and more glucagons. That and the small storage of carbohydrates in the blood soon starts to release fatty acids from the fat deposits and transporting them to the liver where they are metabolized. That leads to increase in ketone production, thereby putting the body in the state called ketosis.

Consistency

You see, getting in better shape will always start with your mind. It doesn't matter what diet you go on, or which exercise program you do. If your mind isn't right, then your waistline ain't getting tight!

How you think, how you respond, and what you desire most will depict whether or not you'll be successful with your program. Therefore, as you can see, focusing heavily on getting your mind right is most certainly the most important step in the beginning.

But how do you do this? Should you just tell yourself "enough is enough... I'm going to get in shape no matter what "and then you just go jump right in with both feet with a diet and exercise program? Do you just mentally block those annoying craving urges that pop up?

Do you just force yourself to put in the exercise DVD, or put your exercise gear on and head to the gym, etc.? Do you just deprive and deny yourself of foods and drinks you can't seem to let go of?

Or...

Do you play it smart and allow your mind to catch up to what your body desires most?

I don't know about you, but that second option sure does sound a whole lot better!

The bottom line is that the most important thing you should never do if you want to stay consistent with dieting and exercising is to never force yourself to start living healthier!

My friend, what I'm getting at here is that you should never force yourself to change and adapt to living healthy. What I recommend you do instead is to allow the changes to happen naturally. And by "naturally", I'm talking about you getting to the point where you no longer desire bad things or prefer to be lazy. Instead, you end up desiring healthier foods and having a more active lifestyle. You will be more willing to stick to your diet if you naturally start despising unhealthy foods and desire more healthy foods. Forcing yourself is a surefire way to end up failing.

It's just like a personality trait. You can't force someone to change how they are, if a change is going to happen, it's going to happen on its own (through experiences, etc.).

Here are a few tips to help explain this a little better:

1. For dieting, yes, I do recommend you get yourself on a reputable diet program. However, to ensure you stay consistent with it, I recommend that you get mind right first.

So, what I suggest you do is to "break into" healthy eating gradually until you start to desire more healthy foods than you do bad foods. In other words, instead of eliminating bad foods, introduce healthier foods into your diet.

In no time you'll start to desire the healthier foods. The reasons why is because you'll begin to see results for one, and secondly, you will discover how better you'll feel after eating healthy foods, and how crappy you feel after eating bad foods or drinking bad beverages.

2. Another example is with water. You don't have to start drinking a ton of water right away. Gradually increase the amount of water you drink each day until you start to desire it more than bad drinks (soda, sugary juices, etc.).

3. For exercising, you do not have to do full insane workouts everyday if you feel like you can't get motivated to them. Instead, you could do quick 1 to 2 minute workouts throughout the day. After some time, you will start to desire doing a full exercising routine. You could literally just stand up right now and do 10 jumping jacks and 5 body squats!

So, if you want to stay consistent with dieting and exercising, start with getting your mind on board. Gradually move into healthy living and before you know it, you'll actually despise bad foods and you'll actually get frustrated if anything is getting in the way of you doing your exercise routine for the day!

Simple and complex carbohydrates

Simple and complex carbs are basically carbohydrates made out of different types and amounts of sugars. The Questions I have been asked a lot is what is the difference between simple and complex carbs and why do we need them...

A carbohydrate is basically one or more of several types of sugar, not the sugar we all know. There are many types of sugars here are some... dextrose, fructose, maltose, glucose and many more.

The only form of sugar your body uses for energy is glucose it`s the simplest sugar of all... no matter What type of sugars the carb contains your body will break it down to glucose which it then uses to fuel various function your body carries out.

When it comes to fitness glucose fuels demanding activities like running and bodybuilding. So the difference between simple and complex carbs is how many sugars they contain simple carbohydrates contain less and complex carbohydrates contain more.

They are called monosaccharide, disaccharides, and polysaccharides.

Monosaccharide contains just one sugar and the best example here would be glucose which remember is the only form of sugar your body can use as fuel... disaccharide contain just two sugars and a good example here would be table sugar you use for tea and coffees.. Because table sugar is made of fructose and glucose.

The most complex are the polysaccharides. They are also called starches and they mainly occur from plants... so foods that are directly derived from plants, for example breads pasta seeds, potato and so on contain a lot of complex carbohydrates..

So when you feed your body carbs that are made up of two or more carbohydrates it has to break them down to glucose so the higher the amount and longer these chains of sugars are that the complex carbs are made up of... the longer your body takes to break them down into glucose. In other words simple carbs that contain only one or two sugar break up and are used much quicker by your body...

And complex carbohydrates with more types of sugars take much longer to break down... so complex carbohydrates make you less hungry through the day gradually releasing energy for longer through the day because they are in your body much longer.

But don't go thinking i am going to just eat all complex carbs because i want to fight hunger for longer because whatever amount of sugars some carbs have they still all differ in the amount of needed nutrients and vitamins they have. So you have to filter this in when trying to eat the right diet... because not all simple carbs are necessarily bad and not all complex carbs are necessarily good.

This is because our foods are more than just sources for carbohydrates... for example fruit often contains many simple carbohydrates and are digested quicker... but fruits also contain vitamins and nutrients that your body needs and benefits from.

On the other hand white bread contains many complex carbohydrates but it is white because the flour it has been made of has been put through a bleaching process that removes nearly all the nutrients that are in flour and just leaves the complex carbs. So in other words not all simple carbohydrates are bad and not all complex carbohydrates are good. It all depends where you get the different types of carbohydrates from.

How does carb cycling really work?

Carb cycling is the basis of many weight loss methods and used in various forms by people who wish to lose fat, build muscles, get toned, and for many more fitness goals. But how does carb cycling work and why is it such a popular method?

Carbs make up the majority of calories we eat. Carbs are found in anything from sugar, through bread and pasta, to fruits and vegetables. There is no shortage of carbs around us. The problem is how to properly eat carbs to burn body fat faster.

This is where carb cycling comes in. Basically, you want to make sure that your body makes the best use of the carbs you consume. By best use I mean provide you with the energy you need to function properly and to workout without gaining weight or body fat.

The problem is that our body pays close attention to how we feed it. When we try to lose fat by cutting down calories or carbs, it immediately begins to worry whether food is about to get scarce. This leads to a reduction in metabolism and a slower fat burning rate, just the opposite of what you want to happen.

This is where carb cycling comes in. Carb cycling (also known as carb rotation) is done by switching up the amount of carbs that you eat from one day to the next. What you're basically trying to do is to "trick" your body so it never knows exactly how you're going to feed it.

For instance, you may cause your body to raise its metabolism in the expectation of a large amount of carbs but then you only feed it a little. This causes your body to burn a lot of calories even though you're only putting in a small amount.

This is just one of the tricks involved, as there are many. The key is to have low, medium, and high carb days to keep your body "guessing" and your fat burning rate as high as possible.

Carb cycling is popular because it produces results and it is a sustainable and doable fat loss method. Indeed, it can even be used to develop bigger muscle mass as it is a system that can be fine-tuned for many fitness goals.

If you find yourself trying diet plans over and over again and failing to lose weight in a consistent manner, you should consider cycling your carbs. It may be just what you need.

IMPORTANCE OF PROTEIN in the carb cycling diet

There are many carb cycling meal plans out there, however, not all of them are going to have the foods that you like on them. So, maybe it's time to create one of your own. Here I will show you what types of foods, and why they should be going into your meal plans.

The carb cycling meal plans are best done on a 5 meal a day plan, this way you will have enough food to fuel you, and keep your metabolism up, which is very important in this diet.

THERE ARE 3 BASIC DAYS in the carb cycling diet:

High Carb Day

Low Carb Day

No Carb Day

On the high carb day, you are going to have 5 meals which will contain as many carbs as you want. Just remember to mix it up with some lean proteins such as:

White chicken/turkey meat

Tuna fish (canned)

Protein shakes with at least 40 - 50g of protein in them. Remember, you will want to go 1-2 g of protein per kg of body weight when you consume protein. This will help maintain your muscle definition.

On the low carb day, you cut back on the carbs to only a limited 3 meals. Just lots of lean proteins, and veggies. Your proteins are very important on this day, as they will be a major part of your diet, so don't skimp out.

On the no carb day you are going to have no to little carb intake. Most of the carbs will come from the veggies, or protein that you will eat throughout the day.

As you can see throughout those 3 basic cycle days, you have to keep up your protein intake. This is what feeds your muscles, without it, you lose them, and your metabolism drops like a rock. Since keeping your metabolism up throughout the diet is the key to making it work then that would mean that eating your proper intake of protein is the key to the whole thing.

Without the proper protein intake at 1-2 g of each kg of body weight, the carb cycling meal plan will be null and void. Because without it, there is no metabolism, because there is no muscle to start it up. So while on the carb cycling diet, please keep up the required intake of protein, this way, you will actually stand a chance in reaching your goal of a nicely defined body.

Chapter 2

Types of Fasting

———

INTERMITTENT FASTING comes in various forms and each may have a specific set of unique benefits. Each form of intermittent fasting has variations in the fasting-to-eating ratio. The benefits and effectiveness of these different protocols may differ on an individual basis and it is important to determine which one is best for you.

Factors that may influence which one to choose include health goals, daily schedule/routine, and current health status. The most common types of IF are alternate day fasting, time-restricted feeding, and modified fasting.

1. Alternate Day Fasting:

This approach involves alternating days of absolutely no calories (from food or beverage) with days of free feeding and eating whatever you want.

This plan has been shown to help with weight loss, improve blood cholesterol and triglyceride (fat) levels, and improve markers for inflammation in the blood.

The main downfall with this form of intermittent fasting is that it is the most difficult to stick with because of the reported hunger during fasting days.

2. Modified Fasting - 5:2 Diet

Modified fasting is a protocol with programmed fasting days, but the fasting days do allow for some food intake. Generally 20-25% of normal calories are allowed to be consumed on fasting days; so if you normally consume 2000 calories on regular eating days, you would be allowed 400-500 calories on fasting days.

The 5:2 part of this diet refers to the ratio of non-fasting to fasting days. So on this regimen you would eat normally for 5 consecutive days, then fast or restrict calories to 20-25% for 2 consecutive days.

This protocol is great for weight loss, body composition, and may also benefit the regulation of blood sugar, lipids, and inflammation. Studies have shown the 5:2 protocol to be effective for weight loss, improve/lower inflammation markers in the blood, and show signs trending improvements in insulin resistance.

In animal studies, this modified fasting 5:2 diet resulted in decreased fat, decreased hunger hormones (leptin), and increased levels of a protein responsible for improvements in fat burning and blood sugar regulation (adiponectin).

The modified 5:2 fasting protocol is easy to follow and has a small number of negative side effects which included hunger, low energy, and some irritability when beginning the program.

Contrary to this however, studies have also noted improvements such as reduced tension, less anger, less fatigue, improvements in self-confidence, and a more positive mood.

3. Time-Restricted Feeding:

If you know anyone that has said they are doing intermittent fasting, odds are it is in the form of time-restricted feeding. This is a type of intermittent fasting that is used daily and it involves only consuming calories during a small portion of the day and fasting for the remainder.

Daily fasting intervals in time-restricted feeding may range from 12-24 hours, with the most common method being 16/8 (fasting for 16 hours, consuming calories for 8 hours). For this protocol the time of day is not important as long as you are fasting for a consecutive period of time and only eating in your allowed time period.

For example, on a 16/8 time-restricted feeding program one person may eat their first meal at 7AM and last meal at 3PM (fast from 3PM-7AM), while another person may eat their first meal at 1PM and last meal at 9PM (fast from 9PM-1PM).

This protocol is meant to be performed every day over long periods of time and is very flexible as long as you are staying within the fasting/eating window(s).

Time-Restricted feeding is one of the most easy to follow methods of intermittent fasting. Using this along with your daily work and sleep schedule may help achieve optimal metabolic function. Time-restricted feeding is a great program to follow for weight loss and body composition improvements as well as some other overall health benefits.

The few human trials that were conducted noted significant reductions in weight, reductions in fasting blood glucose, and improvements in cholesterol with no changes in perceived tension, depression, anger, fatigue, or confusion. Some other preliminary results from animal studies showed time restricted feeding to protect against obesity, high insulin levels, fatty liver disease, and inflammation.

The easy application and promising results of time-restricted feeding could possibly make it an excellent option for weight loss and chronic disease prevention/management. When implementing this protocol it may be good to begin with a lower fasting-to-eating ratio like 12/12 hours and eventually work your way up to 16/8 hours.

Common Question About Intermittent Fasting:

Is there any food or beverage I am allowed to consume while intermittent fasting? Unless you are doing the modified fasting 5:2 diet (mentioned above), you should not be eating or drinking anything that contains calories.

Water, black coffee, and any foods/beverages that do not contain calories are OK to consume during a fasting period. In fact, adequate water intake is essential during IF and some say that drinking black coffee while fasting helps decrease hunger.

If You Just Want The Benefits:

Research on intermittent fasting is in its infancy but it still has huge potential for weight loss and the treatment of some chronic disease.

To recap, here are the possible benefits of intermittent fasting

Shown in Human Studies:

1. Weight loss

2. Improve blood lipid markers like cholesterol

3. Reduce inflammation

4. Reduced stress and improved self confidence

5. Improved mood

Shown in Animal Studies:

1. Decreased Body Fat

2. Decreased levels of the hunger hormone leptin

3. Improve insulin levels

4. Protect against obesity, fatty liver disease, and inflammation

5. Longevity

Intermittent fasting is a feeding pattern which alternates between periods of fasting and controlled eating. It is a simple dietary method divided into many types.

One of the intermittent fasting methods is alternate day fasting, whereby a person takes a normal diet on particular days of the week and fasts on some. During the fasting days, one does not fully abstain from food but rather reduces calorie intake to 1/4 of the normal diet.

The other fasting type is whereby eating is restricted to a certain time window within a day. This means restricting eating between an 8 hour window eating period, which means a person eats once in every eight hours.

Some people however reduce the span to either six, four or even two hours according to their convenience. If practiced accordingly, it can result in a number of positive health effects.

For instance, intermittent fasting promotes general good health. It significantly reduces cravings for snack foods and sugars. The practice normalizes insulin as well as leptin sensitivity. Insulin resistance contributes to many chronic diseases such as diabetes, cancer and heart infections. Intermittent fasting will therefore protect the body from such infections.

Intermittent fasting results in improved brain health. Fasting helps the body to convert stored glycogen into glucose to release energy. If the fasting proceeds for some time, continued breakdown of body fats induces the liver to secrete ketone bodies.

These small molecules are by-products of fatty acids synthesis, and the brain can use them as fuel. Research also indicates that exercise and fasting results in genes and other growth factors which are essential in recycling and rejuvenating the brain.

This type of fasting also boosts body fitness and loss of weight. Combined fasting and exercise increases effects of catalysts and cellular factors so that breakdown of glycogen and fats is maximized. Exercising while hungry therefore forces the body to burn stored fats for significant weight loss.

Intermittent fasting will also boost muscle building especially in men. This is because after eating, the energy gained will be used to sustain a workout session. But if training is done while fasting, the body utilizes stored body fats to sustain the exercises.

Eating after the session ensures that the energy gained is utilized in replenishing the body in the best way. This assists the muscles to quickly recover and build up.

In conclusion, intermittent fasting is a healthy practice but it could result in depression to people who cannot fully sustain it. It needs commitment and perseverance to move through the changes in diet, since only consistency will achieve these positive results.

Chapter 3

———

The Importance of Omega 3

YOU NEED BOTH OMEGA 3 and omega 6 essential fatty acids

When a scientific topic is complex it is easy for incomplete information to be conveyed to the public. This is what has happened with the Omega 3 and Omega 6 scientific story, especially as regards the Omega 3 and Omega 6 advice that people have been given.

The advice to consume foods and supplements that contain Omega 3's, and shun those that contain Omega 6's has been based on misinformation. This advice came about because of a lack of knowledge about how these two polyunsaturated, essential fats work together, and because most people eating a modern diet get very little Omega 3's and more Omega 6's.

These Omega polyunsaturated oils are unique, in that they contain Essential Fatty Acids. They are called essential because the body cannot manufacture them, unlike the other fatty acids, monounsaturated and saturated oils, which the body can synthesize from carbohydrates.

We didn't always know that these fats were essential to our health

The essential nature of Omega 6 and 3 oils was discovered in 1929, although researchers back then thought they were only important for growth and skin health and didn't realize the overall importance to all cells in the body and brain. Later, in the 1960's they discovered that serious symptoms of deficiency arose when these unique oils were not consumed in the diet. But only in the last decade has the importance of Omega 3 become very apparent. However forgetting about the role that Omega 6 plays in our health is a big mistake.

We need essential fats because....

Essential fats are a critical component in every single cell membrane that covers every single one of our 60 trillion cells. The essential fats have a unique structure, which allows them to perform specific functions within the cell membrane, whereas saturated and damaged fats cannot perform the same tasks.

Membranes are the working surfaces of all our cells, so if they are deficient in the right kinds of essential fats, they will be unable to function properly, and nutrients will be unable to get into the cells and toxins will be unable to leave. Furthermore, fat is the second largest compound in our body, and makes up to 60% of the weight of our brain, with up to 30% of that 60% being made up of essential fatty acids. The tiny components that reside inside our cells also need these essential fats to work properly, especially the mitochondria, which is the energy-generator of the cell. So, these essential fats are required in every single cell, otherwise your health will be compromised.

Where do Omega 3 oils come from?

These oils are found in flax seeds, as well as dark green leafy vegetables, and although pumpkin, walnuts, soya, and hemp seeds also contain some of these Omega 3 fatty acids, they have more of the Omega 6's, so are classified as Omega 6 oils. The Omega 3's contain the EFA's Alpha Linolenic Acid, also called ALA's. They are called Omega 3 oils because they have a double bond at the third Carbon atom along from the Omega (or right hand end) of the fatty acid molecule. They are five times more sensitive to damage through light, oxygen and heat than the Omega 6's. And the derivatives of this essential fatty acid, DHA and EPA, are 25 times more sensitive to damage than the Omega 6's. DHA and EPA are also found in fatty cold water fish, like salmon, mackerel and herring.

Why don't we get enough Omega 3's?

At the end of the 19th century, the average diet's Omega 6 to Omega 3 ratio was 2:1 - 4:1. The latest research into our modern diets indicate that those eating a typical western diet, consume more in the ratio of 10:1 to 20:1. Some figures are even closer to 50:1.The reason for this is that more warm weather oils, namely the Omega 6's, are consumed now, such as safflower oil, sunflower oils, corn oil, and products that contain these oils, such as margarine, mayonnaise, peanut butter, ready-made meals, and other mass-produced convenience, processed foods. This is because farmers realized that it was a lot easier to grow these warm weather crops which include Omega 6 oils, about 60 years ago.

Where do Omega 6 oils come from?

Oils rich in Omega 6's include safflower seeds, sunflower seeds, hemp seeds, soya beans, as well as pumpkin and walnut seeds. They contain the EFA's called Alpha Linoleic Acid, sometimes referred to as LA's. These oils are called Omega 6 oils because they have a double bond at the sixth Carbon atom along from the Omega end (or right hand side) of the fatty acid molecule. Most people, estimates are 95%, are getting too much Omega 6 simply because it has become the most popular oil to use in food processing.

So, does that mean we are getting enough Omega 6's?

No, we are not, because the Omega 6's that we are eating today have undergone extensive, cost effective, damaging processing to get them shelf stable. And during this careless processing, these delicate fats are damaged, because they do not like being exposed to heat, light or oxygen. So, although most people are getting a lot of Omega 6's, they get damaged Omega 6's, which can't work effectively in our bodies, because they are damaged. And many of the products that contain these essential fats are partially hydrogenated, which means they are even more stable, but much more damaged, containing trans fats. When people don't know that Omega 6's are damaged through processing, they mistakenly believe that we are getting enough in our diet.

Is taking fish oil alone, the solution?

No, fish oil is not the complete solution for your Essential Fatty Acids (EFA's) needs, for a number of different reasons.

Firstly, fish oil is Omega unbalanced, which means that it only contains the important Omega 3 derivatives (DHA and EPA), but no Omega 6.

This means that a diet supplemented only with fish oil would still lack the necessary undamaged Omega 6's, which are also essential, as described above.

Secondly, nearly all fish oils contain toxins, such as PCB's and heavy metals, like mercury. In the process of trying to remove these dangerous toxic compounds, the fish oil has to be heated to very high temperatures, subsequently damaging the delicate Omega 3 derivatives, DHA and EPA.

Thirdly, fish oil is not as efficient as a plant based Omega 3 supplement, such as flax seed oil, because fish oil only contains about 50% of the Omega 3 derivatives (DHA and EPA) that flax seed oil does. Fish oil is also exposed to light, heat and oxygen during the extraction process, which leads to the damage of these delicate oil molecules. So, you would have to take double the amount of fish oil to get the same amount of Omega 3 as compared to a plant based Omega 3 product, which also means a higher intake of toxins as described above.

Therefore, taking a fish oil supplement does not supply all your Omega needs. You also need to take a supplement that contains undamaged Omega 6's.

Is flax oil alone the solution?

Unfortunately not, because although an Omega 3 deficiency will be cleared up by using flax oil, generally in about 2 - 8 months, a deficiency in Omega 6 will then present itself. This is simply because the body needs both of these essential fats, because they work together in a unique way. If you only use flax oil, the Omega 3 crowds out the Omega 6 oils that you may be getting elsewhere in your diet, and leads to an Omega 6 deficiency, which is indicated by skipped heart beat, fragile, thin, slow-to-heal skin as well as a lowered immune system and painful finger joints, with the comprehensive list of deficiencies found below, under 'signs of having too little Omega 6'

Signs of having too little Omega 3

When you have too much Omega 6, you will end up having a deficiency in Omega 3, because of the special balancing relationship these two essential fats have. The following list is not comprehensive, but will highlight the main deficiencies:

Learning and growth problems

Behavioural and mood changes

Poor muscle growth and muscle weakness

Allergies

Water retention

Dry or irritable skin

Low metabolic rate

Leaky gut

High blood pressure and high triglycerides

Insulin resistance

Tingling in arms and legs

Inflammation in tissues

Depression

Poor motor coordination

Signs of having too little Omega 6

When you have too much Omega 3, which some people are consuming exclusively, if they are only supplementing with fish oil or flax oil, you will end up with an Omega 6 deficiency. Although the following list is not comprehensive either, it highlights the main deficiencies:

Hair loss

Eczema-like skin disorders

Behavioral changes

Growth retardation

Lowered immunity

Failure of wounds to heal

Heartbeat abnormalities

Dry, thin skin, dry hair, as well as brittle nails

Dry eyes

Sterility in males

Miscarriage in females

Kidney malfunction

Arthritis-like conditions

Fatty infiltration of the liver

EMERGING BENEFITS OF omega-3

With a global worth estimated at$8 billion, the market for Omega 3 fatty acids continues to grow as more scientific evidence bolstering a myriad of potential benefits for health is made public. The research being published stresses the importance of Omega-3 fatty acids not only for maintaining overall heart health but also for reducing the risk of cardiovascular disease. Additionally, Omega-3 is thought to play a crucial role in improving eye and brain development in the formative years of life (prenatal and early childhood) and in the cognitive brain functions of the elderly.

Emerging studies on Omega-3 fatty acids continue to reveal new benefits that go well beyond supporting heart and brain health. Scientists now believe that the two most important Omega-3 fatty acids- eicosapentaenoic acid (EPA) and docosahexaenoic acid (DHA)-may play a prophylactic role in almost every human disorder.

Emerging roles of omega-3 fatty acids

Cancer Potential

New evidence from research suggests a role for Omega-3 in the prevention and therapy of certain types of cancers. In Japan, epidemiological studies showed an increase in the incidence of breast cancer in people who shifted from the traditional Japanese diet consisting primarily of fish, to a more westernized diet and lifestyle. Since then, several studies have confirmed an association between the consumption of Omega-3s with a decreased risk of breast, prostate, colon and kidney cancers. Scientists theorize that Omega-3s may actually work to oppose the proliferation of cancer cells, but more research is needed to know just how the components of EPA and DHA work together and which ones play a crucial role. Studies also suggest that Omega-3 supplements may work in synergy with chemotherapeutic drugs to help reverse the side effects of chemotherapy, as well as prolong patient life.

POTENTIAL PROTECTION against inflammation

Reducing neural inflammation and increasing anti-apoptotic mediators, Omega-3 fatty acids exhibit potential benefits for professional athletes engaged in contact sports where head injuries typically occur. Statistics show that repeated sports-related concussions may result in higher risks of depression and mild cognitive impairment which is a precursor of Alzheimer's disease. Initial studies conducted by the West Virginia School of Medicine, shows that DHA supplementation may help to reduce the production of amyloid precursor protein (APP) following traumatic brain injuries. It was also seen to reduce bio-markers associated with neural inflammations and cell death (apoptosis).

The anti-inflammatory properties of Omega-3s have shown benefits for people who suffer from joint discomforts by reducing joint stiffness and

pain, increasing grip strength and improving walking pace in people who
have arthritis.In some cases, supplementation is associated with lowered use
of anti-inflammatory drugs (NSAIDs).

A number of clinical trials have assessed the benefits of Omega-3 fatty acids
in several inflammatory and autoimmune diseases including rheumatoid
arthritis, Crohn's disease, ulcerative colitis, psoriasis, lupus erythematosus,
multiple sclerosis and migraine headaches.

Potential immune booster

Expectant mothers taking 400 mg of DHA were seen to have babies with
stronger immune protection against colds than babies of mothers who did
not take any DHA supplementation, according to the Journal of Pediatrics.
Data from the study shows at one month, babies whose mothers took DHA
supplementation were seen to have experienced 26%, 15% and 30% shorter
duration of cough, phlegm and wheezing respectively. At six months of age,
DHA babies had 20 % shorter duration of suffering from fever, 13% shorter
duration with a runny nose, and 54 % shorter duration with breathing
difficulties.

————————————

POTENTIAL BENEFITS against cellular aging

People diagnosed with heart disease and had high levels of DHA in their
blood were seen to have longer telomeres than those with low levels of DHA
in their blood. Telomeres are the protective ends of chromosomes. Scientists
have long observed the association between the length of telomeres and
cellular aging. When telomeres become short, the cell is no longer able to
reproduce itself and becomes apoptotic. Shorter telomeres are associated
with higher cardiac mortality and other undesirable consequences of aging.

Omega-3s may help reduce the risk of degenerative muscle loss in the elderly.
The American Journal of Clinical Nutrition (2011) reported a 400 mg daily
dose of Omega-3s was seen to increase the rate of muscle protein synthesis,
thereby increasing the production of muscle protein. The eight-week study

also observed an increase in the supply of amino acids and insulin in the elderly.

Potential mood stabilizer

A new study published in Brain, Behavior and Immunity reports a reduction of anxiety and inflammation fromOmega-3 supplementation. Cytokines are compounds that promote inflammation. Cytokine production is known to increase with psychological stress. The study, conducted by researchers of Ohio State University, studied stress levels and cytokine levels in a group of young, healthy medical students appearing for their medical exams. Those receiving Omega-3 supplements showed a 20 percent reduction in anxiety compared to the placebo group. The study conducted also showed a reduction in the level of cytokines in the body, thus showing the potentially positive influence of Omega-3s on anxiety and inflammation.

Sources of Omega-3 fatty acids

Look for high-quality Omega-3 manufacturers that have a wide range of source options such as:

Krill (rich in phospholipids, Omega 3 fatty acids and an antioxidant called astaxanthin), is found abundantly in seawater-up to 30,000 creatures in one cubic meter.

Algae-sourced Omega 3 fatty acids, an environment-friendly, vegetarian source of Omega-3s.

Green lipped mussel, an eco-friendly seafood rich in Omega-3s.

Plant-based sources such as flax seeds that contain alpha-linolenic acid (ALA) are popular but concerns conversion ratio of ALA into DHA and EPA is considered to be poor. However, ongoing research on bio-engineering plant-based sources of Omega-3 from canola and soy is leading to greater, innovative delivery options for Omega-3 manufacturing.

Chapter 4

———

Basics of Protein

———

THE TOPIC OF PROTEIN is of primary interest to all in the fitness industry these days. Many celebrities and athletes give praise to high protein diets for making their lean, sexy bodies possible.

High protein food has become the overall favorite weight loss food group due to its effect on body metabolism and constitution, and is no longer exclusive to the bodybuilding/weightlifting sector.

The popularity for high protein intake has been incited by the several diets and books written that promote the advantages of higher dietary protein and essential fats, lower refined carbohydrates and very low saturated fats.

Protein has also been touted as a mood stabilizer plus control for diabetes, but critics point out that restricted intake of calories endures as a factor regarding successful weight loss.

The word protein, derived from the Greek word "protos"meaning to come first, is true to its meaning. It is vital to almost 300 billion cells that are being built each day in our bodies.

If we do not eat enough protein, we are then lacking the very basic building blocks necessary to produce hormones and enzymes, components of the immune system, and primary structure of nearly all body tissues.

How much do I need?

The minimum daily protein requirements as set by WHO (world health organization) are 0.8 grams per each kilogram of body weight, however individual protein needs are dependent on one's level of activity, overall health status, and age.

Those recovering from injury require extra protein for tissue healing. Athletes training hard to gain muscles or who may be causing small injuries also need extra for growth and repair.

Protein needs in children vary from infancy through to late adolescence, with large amounts of new protein tissue being created during rapid growth spurts. The protein requirements then stabilize once early adulthood is reached. The following table gives the RDA (recommended dietary allowance) of protein throughout the lifespan:

Age in years grams/kg of bodyweight grams/pound of bodyweight

0-6months 1.52.69

6mos-1 year 1.10.50

1-3 years 1.10.50

4-8 years 0.95.43

9-13 years 0.95.43

14-18 years 0.85.39

19 & older 0.8.36

Proteins are chemical complexes incorporating hydrogen, carbon and oxygen, the same as fats and carbohydrates do, but they also contain roughly 16% nitrogen. These four elements combine in various ways (on occasion also with sulfur) to form amino acids which are the building blocks of proteins.

Twenty amino acids, are used in varying combinations to make up the proteins needed for the human body. However, the body is not capable of synthesizing nine of these amino acids. These nine are known as "essential" or "indispensable" amino acids and include:

- Histidine

- Isoleucine

- Leucine

- Lysine

- Methionine

- Phenylanine

- Threonine

- Tryptophan

- Valine

The non essential list consists of Alanine, Arginine, Asparagine, Aspartic acid, Cysteine, Glutamic acid, Glutamine, Glycine, Proline, Serine and Tyrosine.

What foods are high in protein?

Foods that contain adequate amounts of all nine essential amino acids are known as complete proteins, whereas those missing or having inadequate amounts of one or more of the essential amino acids are called incomplete proteins.

Amino acids found in animal and plant sources are of the same quality, but animal proteins are complete because they contain each essential amino acid in the appropriate amounts that the body requires.

This group contains meats, fish, poultry, eggs, milk and dairy. Plant proteins are less concentrated than animal proteins and are incomplete, but when eaten in certain combinations they can provide all the amino acids required for complete proteins.

Legumes, like soybeans, split peas, garbanzo, kidney, lima and black beans to name a few, are good protein sources. They are also high in carbohydrates so are normally listed as a starch. Nuts are high in protein, but also in fats. Grains are another group of plants that have protein content.

Grains when combined with legumes (such as beans and rice) often supply all the amino acids needed for complete proteins. Animal foods are also fine sources of vital minerals like calcium, iron and zinc whereas plant foods supply carbohydrates, fiber and phytochemicals.

Are there health risks to excessive protein?

There has been no set upper daily limit for protein, but continual high doses have been connected with some chronic diseases.

It can cause stress to the liver and kidneys from the processes required to excrete excess nitrogen from the body, plus the increase in urine acidity leads to increased calcium excretion. Increased calcium in the urine can promote formation of kidney stones.

Increased excretion of calcium can also lead to weakening of bones and joints, while the formation of uric acid during metabolism of proteins can accumulate in joints causing inflammation and exacerbation of conditions like gout. Ensuring adequate calcium intake helps to defray some of these effects.

Under normal circumstances, we can obtain plenty of good quality protein from our diet. Supplementation is a good consideration, though, if one is stressing the body through training or healing, and requires protein in particular in higher amounts.

What protein does for you

All the food you eat can be divided into one of three macronutrients: fats, carbohydrates, and protein. Protein is different from the others, because it is more likely to be burned as energy or used for other purposes, rather than being stored in the body.

Fat is much easier to digest, while carbohydrates can vary depending upon type. Simple carbs digest quickly and easily, causing sugar spikes that lead to weight gain. Complex carbs take a bit longer to digest. Their more even digestive rate does not cause sugar spikes.

Protein takes a while to digest, but it starts to burn energy the instant it finally enters the bloodstream. This macronutrient is of vital importance to every cell in the human body. While it is necessary for life, you can eat too much of it, just as with any nutrient.

People need protein the most during the first six months of life, when according to body weight, an infant needs double the amount of protein an adult, or even an older child will need.

What exactly does protein do?

It would be impossible to list everything protein does here, but here are some of the things that make protein necessary.

Building and repairing muscle

Building connective tissue

Adding material to the bone matrix

Adjusting the pH balance of the blood

Helping to form certain hormones and enzymes, such as those that regulate sleep, digestion, and ovulation

Strengthening the immune system (antibodies are made from protein)

Creating new blood cells

Forming RNA and DNA

Making new neurotransmitters

How Protein Is Digested

The building blocks of protein are called amino acids. Most of these amino acids can be manufactured by the body without any help, but there are eight of these which the body cannot synthesize. These must be provided by food on a daily basis.

These eight are known as the essential amino acids. A protein that has all of the essential amino acids is a complete protein, while those that lack even one are incomplete proteins. Proteins that come from animal sources are complete proteins, while almost all plant-based proteins are incomplete.

The amino acids, in turn, are simple compounds made from carbon, hydrogen, oxygen, or nitrogen. These amino acids link themselves into chains known as peptides, which can have more than 500 amino acids within them.

The protein you eat is broken down into these basic amino acids when your body digests them, so they can be used to create new amino acids and certain enzymes and hormones as your body absorbs them.

Once protein enters the stomach, hydrochloric and gastric acid reduces it down to its basic components. There is an enzyme in the stomach known as pepsin, the only enzyme able to digest collagen (a protein in the connective tissue of animals), which digests the amino acids.

These acids then move to the duodenum, the first section of the small intestine. More enzymes act here, breaking down amino acids into even smaller portions until they are small enough to pass through the lining of the intestines and directly into the bloodstream.

Exercise reduces your body's production of protein. The protein that remains is converted into energy that enables your muscles to continue working. Once exercise is finished, the rate of protein production remains at a low for about twenty-four hours, while the burning of energy continues to be high.

This is especially true for heavy resistance training. If no new protein is consumed during this period, the breakdown rate will be greater than the synthesis rate and the body will start taking fuel from the muscles.

How proteins are evaluated

Proteins can be evaluated for their value, especially by endurance and strength athletes who rely upon protein for their performance. Such people usually judge protein on two scales. The first is the Protein Digestibility

Corrected Amino Acid Score, or PDCAA, which evaluates a protein on its completeness, which a complete protein rated as a 1.

The second evaluation score is the biological value, or BV, which is determined by how much of a particular protein is retained by the body once it is broken down into its basic components.

Both of these scales base their standard upon the egg - an egg is a complete protein and 100% of it is retained by the body. It is unlikely that anyone but an elite athlete will care about the PDCAA or BV of any given protein. Luckily, there are simpler ways to determine how good a protein might be for you.

First, your proteins should be low in fat, especially saturated fat. They should also be low calorie in relation to the size of the portion. They should also have other nutrients that are important to your day-to-day life. Good tasting is another priority. If it doesn't taste good, you're not going to want to eat it.

Any protein supplements you use should be high in protein, of course, but pick the ones that are low in calories, without any added fats or sugars. You should look at the labels of any supplements you buy, especially protein bars. Many of them claim to be health bars but are not much different than the standard candy bar.

How much protein is necessary?

Everyone needs protein, no matter what their age or circumstances. Just how much is needed varies with age, health, weight, and activity level. Protein is like any other nutrient in that it should not be taken to excess.

Anything more than 35% of all calories per day is too much, even for athletes who need a high supply of protein. During the first six months, a baby needs about 2.2 grams of protein per kilogram of body weight, but they are the only ones who should ever consume that much.

Even the most dedicated bodybuilder should have no more than 1.6 grams of protein per kilogram of bodyweight. For the average person, 0.8 grams per kilogram of body weight is enough. That's about 60 grams of protein per day for the average man, which is about 220 grams of meat.

Women can get by on less protein than men, unless they are pregnant, in which case they will need more protein. Too much protein for a non-pregnant woman can lead to a loss of calcium through the urine, which can in turn increase the risk of osteoporosis.

Frequency of protein consumption

The majority of like-minded bodybuilders like you and me usually follow the 2-3 hour rule when it comes to the 5 meals we must consume every day. Whether we are on a fat loss or bulking diet, we must always consume complex meals every time (Protein, Carbs and Fats). A larger portion of proteins for each meal is always necessary, no matter what our aspirations are.

When it comes to carbs and fats, we may have to change things up a little according to our short-term goals. We do this 5 times a day with a few protein shakes here and there; and that is usually it. With this practice only, we have put ourselves on the road to more muscle mass or less body fat, depending on our overall goals.

Where does the myth of periodic food intake come from?

There is no way of knowing. It once became a basic rule, and now everyone is blindly following it. We usually justify this theory by claiming that our digestive cycle needs approximately 3-4 hours to digest the protein.

Also, if we do not follow this general myth, our muscles would not get the necessary nutrients in time for ensuring proper muscle building mechanisms - at least this is the overall belief.

In many cases, for example, if we are following a rice-chicken-based meal plan during the day, it might take longer for our stomach to digest the food we eat. A larger portion of scrambled eggs may sit 3-4 hours in your stomach, and it still would not get digested.

Then, there is cottage cheese: the casein in it is also hard to digest. What happens when we have a meal that takes our stomach more time to digest? Do we really have to follow the 2-3- hour rule in this case as well?

Quantity is as important as frequency when it comes to protein consumption. There are certain regenerative anabolic processes that start up shortly after a meal, and the extent to which they are successfully completed are based on how much protein we take as well as how long these proteins ensure a proper stream of amino acids for our muscles.

Chapter 5

———

Liquid Diets for Body Detoxification

WHEN GASTROINTESTINAL illnesses occur, or when weight gain becomes apparent, or an upcoming medical procedure is forthcoming, liquid diets are sometimes used to prepare the body for wellness and treatment.

Usually, when following such specialized diets, individuals will avoid consuming solid foods or pursue a more restrictive approach towards their intake of liquid sustenance. Liquid diets are often suggested or required by doctors before various surgical procedures or embraced by overweight people who wish to jumpstart a diet and exercise plan.

When considering this type of diet that excludes solid foods, you will find that there are two main types: clear and full.

LIQUID DIET CHOICES

Clear liquid diets consist of specially prepared and visually transparent foods, such as meat and vegetable broths, bouillon, clear fruit juices, clear fruit ices, black tea, black coffee, popsicles, and clear gelatin desserts.

The purpose of this type of diet is to maintain essential bodily fluids, minerals, and salts. Clear liquid diets also provide energy to people when their normal food intake is disrupted.

Since the body easily absorbs liquefied foods, the digestive system undergoes less stress, while unwanted chemicals, toxins, and wastes are flushed from the intestinal tract. This is why clear liquids diets are often associated with the preparation for surgery.

The gentleness on the body of such a diet is also suggested for a first day of eating and drinking after surgical procedures and before medical tests, such as a colonoscopy, or certain x-rays.

Full liquid diets consist of both clear and opaque foods (with a smooth consistency). One may consume milkshakes, ice cream, milk, strained cream soups, oatmeal, fruit nectars, pudding, and honey. Sometimes, such a type of diet may allow other kinds of foods that are thinned and blended with fluidized nutrients (like fruit juices or milk).

The purpose of a full liquid diet is to make a proper transition from a clear liquid diet to a regular routine. This also helps people ease into normal eating patterns, especially after a session of fasting or undergoing surgery. Certain procedures and medical concerns also respond well to this particular type of diet, such as jaw wiring and problems regarding swallowing and/or chewing.

Pros and cons of liquid diets

The advantages associated with liquid diets involve rapid results and less strain on the body. This type of diet is also one of the easiest methods to begin and follow. Despite the ease and beneficial results, this restrictive diet and nutrition regimen has also been known to bring about an assortment of disadvantages.

Today, it is not uncommon for people to consider a liquid diet when trying to lose weight, as such diets have been reported to have helped individuals shed between two and twenty pounds. Typically, however, this type of diet is often recommended for the overweight or obese.

Additionally, patients facing bariatric surgery are advised to follow a fluid-only diet for up to 10 days after their procedure. Interested parties may safely integrate a liquid diet plan into their lifestyle by replacing solid food in two out of three meals with the specialized liquid replacement diet.

People who are in need of losing a large amount of weight in order to enhance their health often benefit from such diets, but should approach the process as a short-term solution.

Full liquid diets are dangerously low in vital vitamins, such as iron, vitamin B12, and vitamin A.

Clear liquid diets are drastically lower in calories and important nutrients. For example, many of such diets discourage meat and grains.

Overall, if you are following this particular type of diet and nutrition plan, you should keep tabs of your nutritional needs and confine the dieting period to a limited time frame. Also, it may be beneficial to engage a nutritionist to help monitor your diet and nutrition needs during this dieting period.

Liquid vitamin nutrition is your key to good health

High quality liquid vitamins and nutritional supplementation is essential to make up the deficit of nutrients in our food. The transition from pill form to liquid vitamins and minerals is taking place and the wave of the future for the health industry and nutritional supplements is liquid vitamins.

These vitamins are digested as they pass through the intestinal tracts and the step of breaking down the pills is eliminated. Liquid Vitamins are much better absorbed than pill vitamins while pills are more taxing on your digestive system and much harder to swallow!

They also help you achieve better health much faster than pill form. Maximum absorption is always key when it comes to liquid vitamins. Why should you use a liquid supplement?

Instead of pills, which sometimes don't digest, the liquid is absorbed into the system fully. The main reason why liquid vitamins are more effective than pills and tablets is due to the nature of their liquid base.

Medical studies have shown that pills and capsules only have a 25%-40% absorption rate at the cellular level, while liquid nutritional products exceed a 90% absorption rate. This is true especially in the case of old people and children, for whom liquid vitamins are available in a variety of flavors.

Feed your 300 million new cells the nutrition they need to thrive with a complete good tasting and powerful liquid vitamin available. A person with acid-refux disease is going to have a harder time with liquid vitamins because of the high concentration of acid built up.

These liquid dietary supplements are high potency formulas you can count on. Multiples (in liquid form) give your body more daily vitamins and minerals through whole food vegetation (than pills).

Today's health and nutrition is not what it used to be and is headed nowhere fast. Whether you are in good health or not, proper nutrition is essential for good health. Health is a major concern worldwide, from vitamins and minerals, to the proper nutrition and intake, it is hard to get the proper amount of nutrition you need everyday from just the regular foods you intake.

Like all of us, we take vitamins with the intention of getting good nutrition, but when they are pushed out of our system virtually undigested then that creates a problem. With the fast paced world we live in today, it's hard to get the right vitamins, minerals, and nutrition into our body everyday. The body's cells are always starved for proper nutrition.

Supplementation of liquid vitamin nutrition is the secret to protecting you and your family against disease, staying healthy and looking and feeling youthful.

How water helps to lose weight and liquid diets

In the liquid weight loss diet you would need to intake any form of liquid. It is known that water helps to lose weight. Food of any kind in a liquid form can be taken in this particular diet. The liquid can be opaque or clear in nature.

There are two types of liquid diet that you can follow for weight loss. They are clear liquid and full liquid diet. Water helps to lose weight with both advantages and disadvantages.

There are a number of differences between the two diets. In the clear liquid diet you can intake clear or transparent foods such as clear fruit juices, vegetable broth, and meat broth etc. where as in the full liquid diet you can only intake liquids which is totally transparent or opaque. The most important thing to keep in mind where water helps to lose weight is that the liquid should have a smooth consistency.

The doctors prescribe liquid diets to the patients who undergoing medical test or any kind of surgery. This form of diet is given to patient who wants their stomach to be devoid of any kind of solid food particles.

On the other hand people who suffer from ailments like vomiting or diarrhea, or for that matter have chewing or swallowing problems are prescribed a liquid weight loss diet in the event of restoring the digestive system. In fact there are many of us who go in for liquid diets to just lose weight.

There are many advantages that one can get. They are:

This form of diet is useful for you when your body needs detoxification and at the same time your digestive system needs to be restored. If the liquid diet is followed in such a case then it drains all the toxins and wastes from the body.

This form of diet is most successful for medical purpose.

Some of the liquid diets that you follow are so tasty and filling that the craving that you have is sustained.

Some liquid diets also emphasize on consumption of raw materials. With this you can avoid sweets, saturated fats and other fats. On the other hand you are eating and drinking food that is nutritional for the body.

A liquid weight loss diet not only detoxifies the body, but it also rejuvenates the body. The most important thing to keep in mind is that water helps to lose weight but you should not overdo the diet. Otherwise it might be harmful for you.

Chapter 6

———

The Functions of Insulin

INSULIN IS A NATURAL hormone controlling the level of sugar glucose in the blood. Insulin helps cell use glucose for energy. Cells cannot utilize glucose without insulin. When excess glucose produces in the bloodstream, it increases the risk of diabetes.

The primary source of the fuel is insulin. Insulin let body cells take glucose from the bloodstreams. The cells may use glucose for energy production if it is required. Alternatively, the glucose is sent to the liver for preservation in the form of glycogen.

To define insulin broadly, it can be added that insulin is a hormone, a central regulating and glucose metabolism in the body. Insulin helps cells in the liver; muscle and fat tissue take glucose from blood. It is stored in the liver and muscle as glycogen.

Insulin prevents to use fat as an energy source. When insulin is absent, body cell does not take glucose, and body resumes utilizing fat as an energy source. When Adipose tissue transfers lipids to the liver, it mobilizes as an energy source.

When control level of insulin fails, this failure will result diabetes mellitus. Thus, insulin is used to treat some forms of diabetes mellitus medically.

Type 1 diabetics depend on external insulin injecting subcutaneously to survive as because the system is not producing hormone no longer.

Type 2 diabetics are insulin resistant, this resistant perhaps suffer from insulin deficiency. It appears patients with Type 2 diabetes may eventually require insulin if other medications fail to control blood glucose levels at a certain extent although this is somewhat uncommon.

Insulin also influences other body functions including vascular compliance and cognition. Once insulin enters the human brain, it strengthens learning, memory and it particularly benefits verbal memory.

Insulin regulates glucose metabolism and deals with stimulates lipogenesis, diminishes lipolysis, increasing amino acid transport into cells, modulation of transcription, altering the cell content of numerous mRNAs, growth stimulation, DNA synthesis and cell replication.

Therefore, the elaborated function of Insulin can be countering the concerted action of a number hyperglycemia generating hormones and maintaining low blood glucose levels. Because there are many hyperglycemia hormones, untreated disorders associated with insulin leading to severe hyperglycemia and shortened life span.

Moreover, the function of insulin is critical for having a healthy body. Pancreas produces insulin. Islet cells usually produce it. The pancreas behind the stomach is a vital organ. The pancreas produces all types of digestive enzymes and hormones that are designed to breakdown the food.

Insulin is considered for both a protein and a hormone. It is the regulating body to distribute the necessary amount of blood sugar requiring in each cell. When we eat, the food is converting into glucose. This is simply sugar. In addition, it is widely known as "blood sugar level".

Insulin regulates glucose metabolism, it also stimulates lipogenesis, diminishes lipolysis, and increases amino acid transportation into cells.

Insulin resistance

Insulin is a hormone that is produced by pancreas, which allows cells to use glucose (sugar) as energy. People with insulin resistance have cells that don't use insulin effectively, which means the cells have trouble absorbing glucose.

The diminished ability of cells to respond to the action of insulin in transporting glucose (sugar) from the bloodstream into muscle and other tissues results in a build-up of glucose in the blood. As a result, the body needs higher levels of insulin to help glucose enter cells.

The pancreas tries to keep up with this increased demand for insulin by producing more. As long as it is able to produce enough insulin to overcome the insulin resistance, blood glucose levels stay in the healthy range.

But over time, the pancreas fails to keep up with the increased demand for insulin. Without enough insulin, excess glucose builds up in the bloodstream, leading to pre-diabetes, diabetes, and other serious health disorders.

Symptoms

Insulin resistance usually has no symptoms. People may have the condition for several years without knowing they have it.

People with its severe form may develop dark patches of skin, usually on the back of the neck. Sometimes people have a dark ring around their neck. Dark patches may also appear on elbows, knees, knuckles, and armpits. The skin changes usually appear slowly. The affected skin may also have an odor or itch. This condition is called acanthosis nigricans.

Causes

Although its exact causes are not completely understood, there are certain contributors for it as described below:

Obesity - Many experts believe that obesity is the primary cause of insulin resistance, especially excess fat around the waist. According to many studies, belly fat produces hormones and other substances that can cause serious health problems such as insulin resistance, high blood pressure, imbalanced cholesterol, and cardiovascular disease (CVD).

Belly fat plays a part in developing chronic inflammation in the body. This inflammation can contribute to the development of insulin resistance, type-2 diabetes, and CVD. However, studies show that losing the weight can reduce insulin resistance and prevent or delay type-2 diabetes.

Physical inactivity - Many studies indicate that physical inactivity is also linked with insulin resistance. Studies show that after exercising, muscles become more sensitive to insulin, reversing insulin resistance and lowering blood glucose levels.

Exercise also helps muscles absorb more glucose without the need for insulin. The more muscle a person has, the more glucose one can burn to control blood glucose levels. Therefore, weight training plays a significant role in reversing it as it helps muscles grow.

Sleep problems - Many studies show that sleep problems, especially sleep apnea, can increase the risk of obesity, insulin resistance, and type-2 diabetes. Night shift workers may also be at increased risk for these problems. Sleep problems also result in poor sleep quality producing sleepiness or excessive tiredness during the day.

Other causes - Other causes include ethnicity, certain diseases, hormones, steroid use, some medications, older age and cigarette smoking.

The bottom line

This is a disturbing fact that insulin resistance is common all over the world and its prevalence continues to increase. Another statistically disturbing trend has been observed that one-third of obese children and adolescents have it, who form a high risk group for development of type-2 diabetes and cardiovascular diseases. This becomes all the more significant as there is rising incidence of obesity in children and adolescents worldwide.

It is undeniable that type-2 diabetes has reached epidemic proportions worldwide. Simultaneously, due to high prevalence of overweight and obesity coupled with physical inactivity and unhealthy food habits, there is steeply rising trend of insulin resistance in adults as well as adolescents globally.

This portends to further worsen high prevalence rate of type-2 diabetes. Therefore, it requires that it should be checked in time so as to avoid the situation from going out of control.

Effects of high insulin

Insulin is a hormone produced by the body. It's essential for good health. But too much insulin is harmful to your health and to your fitness goals. Insulin's function is to lower your blood sugar levels when it gets too high. Without proper insulin levels you could develop hyperglycemia or diabetes mellitus.

But too much insulin has negative effects, it:

Promotes fat storage

Inhibits fat burning

Increases your appetite and hunger

Each of these side effects ruins your effort to get healthy and toned.

Avoid excess insulin by monitoring the amount of carbohydrates you consume. In the body, high carbohydrate foods convert glucose and rapidly enter the bloodstream. Glucose in the bloodstream prompts the secretion of insulin and glucose can cause excess insulin.

Insulin shuttles glucose out of your bloodstream. This shuttling action of insulin cuts off the release of fat for energy. Stored fat becomes unavailable for energy (fat burning). Insulin promotes fat storage.

Excess insulin also cuts off the release of the hormone glucagon. Glucagon promotes fat burning by inhibiting fat-storing enzymes. Instead it mobilizes fatty acids from fat stores to be burned for energy.

It is important to know that insulin and glucagon are paired hormones. That means that when insulin is elevated, glucagon is suppressed. When insulin levels fall, glucagon is elevated. So even if you include protein in a high carbohydrate meal, the insulin produced by the carbohydrates will not allow your body to produce a sufficient amount of glucagon.

Eating pure protein only raises glucose a little. This happens whether you're following a low carbohydrate diet or a typical high carbohydrate diet. However, while on a low carbohydrate diet the protein meal produces a very little rise in insulin, but a significant rise in glucagon...the perfect conditions for fat burning!

On the other hand, the protein meal on a high carbohydrate diet produces a large rise in insulin and only a slight rise in glucagon.

There is no such thing as an essential carbohydrate. Haphazard carbohydrate consumption can wreak havoc on the best of intentions. But this doesn't mean carbs should or can be avoided all together.

The brain prefers carbohydrates for fuel because they convert so easily into glucose. They can be used for immediate energy. But know this, there are smart carbs and dumb carbs.

All carbohydrates are not alike

There are simple carbohydrates and complex carbohydrates. Both have very different effects on your blood sugar level and the secretion of insulin. And, remember the dangers of excess insulin release in the body! "So...What's the difference?" You ask. Well...

Simple carbohydrates are sugars. Like yogurt, honey and fat-free cookies. Simple carbs are digested easily and enter the bloodstream rapidly signaling insulin release...thus inhibiting fat burning.

Grab a banana before running out the door to the gym? The banana will be in your bloodstream before you can climb two flights on the stairmaster. Now you're getting stairmaster energy from the banana...and not from fat storage.

Complex Carbohydrates are starches, like bread, potatoes, and rice. Some complex carbs are digested and absorbed into the bloodstream slowly-keeping your blood sugar at a constant level which is the goal! But keep in mind that there are complex carbs that react in the same way as simple carbs.

Complex carbohydrates, like potatoes and regular wheat bread, enter the bloodstream rapidly because the body doesn't have to do much to digest these foods. It's safest to choose carbohydrates that enter the bloodstream slowly...that is, how they place on the Glycemic Index (GI).

Ask any diabetic about the glycemic index and you're sure to get a detailed and concise explanation. The glycemic index (GI) is important to diabetics because it indicates the rise in blood sugar level caused by eating specific carbs. The glycemic index calculates the rate at which a consumed carbohydrate will break down in the body and enter the bloodstream.

Carbohydrates lower on the index are better because they don't produce a blood sugar rush. They digest and enter slowly, keeping blood sugar level. This allows your body to absorb all the nutrients smoothly. These are smart carbohydrates.

The higher a food is on the glycemic index, the faster it enters the bloodstream, and the faster it inhibits your fat-burning ability. These are dumb carbohydrates. Remember, just because a carbohydrate is complex doesn't mean it will have a low glycemic index.

Some complex carbs are high glycemic, like rice cakes, white bread, oats, and carrots. So choose carbohydrates by their glycemic score, not by whether or not they're complex or simple. Achieving and maintaining proper blood sugar metabolism is essential for a lifetime of excellent health.

Impact of High Insulin Levels on Weight Loss

If there is one physiological concept that plays a very important role in an individual's ability to lose weight, it would be the secretion of insulin.

The elevated secretion of insulin and the resultant insensitivity of insulin receptors are some of the major causes of excess bodyweight gain and can also be considered as factors that have made it pretty difficult for several millions to effectively lose weight.

The human body is a marvel when it comes to creating systems of checks and balances to make sure that the body operates optimally.

This process is known as homeostasis and is a complex set of biological regulatory systems employed by the human body to maintain an optimal physiological and chemical equilibrium in order for cellular reactions to occur. Examples of these homeostatic functions include the body's self-regulation of hormone and acid-base levels, the composition of body fluids, and also that of cell growth and body temperature.

When it comes to insulin, it is a hormone which helps the body to regulate the amount of glucose available in the bloodstream. Whenever there is an excess amount of glucose in the blood, the body in a counteracting move secretes insulin which consequently instructs body cells to take up the excess glucose and convert it to either glycogen for storage in liver and muscle cells or fat to be stored in fat cells.

However, it is important to state at this point that besides insulin, other hormones such as glucagon, cortisol, growth hormones, epinephrine, and norepinephrine, also have certain amount of influence on overall blood glucose levels.

While insulin is generally known to help in lowering blood glucose (sugar) levels, virtually all the others work in one way or the other to increase blood glucose levels with glucagon having the most significant effect.

After eating, the amount of glucose (the final product of the digestion of carbohydrate-containing foods) in the bloodstream increases. The amount of glucose that is produced and the rate at which they are absorbed into the bloodstream to cause an increase in blood glucose level is influenced by a number of factors.

The major factors that can significantly influence glucose absorption include the glycemic index of the consumed carbohydrate food and also the co-ingestion of fats and proteins.

Generally, carbohydrate foods with high glycemic index (above 65) are very quickly absorbed into the blood and cause a quick rise in blood glucose level while those with a low glycemic index (below 50) are absorbed slower and cause a gradual rise in blood sugar levels.

In fact, the body is very sensitive to the overall amount of glucose in the bloodstream and works vigorously to ensure its stability. Therefore, when blood glucose levels increase above normal, specialized cells in the pancreas secrete insulin to help remove the excess glucose for storage.

Unfortunately, constantly high level of glucose in the bloodstream can destabilize this delicate system making the body incapable of effectively removing glucose from the bloodstream.

This is due to the fact that when there is a constant high amount of glucose in the bloodstream due to overeating, with time the insulin receptors on the surface of cell membranes (that normally carry out the uptake of glucose from the bloodstream) may become "de-sensitized" or "numbed" to the effect of insulin.

The excess glucose removed from the bloodstream is stored mainly as fat because under normal circumstances the body can only store a certain amount of excess glucose in the muscles and liver cells. When the muscle and liver options become filled up, the body has only one alternative left and that is to store the excess glucose as fat in fat cells (adipose tissues).

Furthermore, excessive insulin secretion does not just stimulate the storage of excess glucose as fat in the body but it also hinders the body from releasing and utilizing fat for use as an energy source. Thus, insulin not only causes the storage of excess body fat but it also stops the body from burning fat.

Therefore, excessive insulin secretion results in insulin resistance which encourages the storage of fat in major fat-storage areas such as the hips, thighs, and stomach. This condition practically makes permanent weight loss almost impossible.

Fortunately, insulin levels can be reduced through engaging in regular physical exercise which has been demonstrated to be capable of making insulin receptors on cell membranes more sensitive to the effect of insulin.

Also, a reduced insulin level can be achieved through a reduction in the overall amount of consumed carbohydrates especially the high glycemic index types. Reducing carbohydrate consumption can be argued to be the best way to go about reducing the amount of glucose in the bloodstream and thereby a reduction in the secretion of insulin.

When blood glucose level is low, the body secretes the hormone glucagon which instructs body fat cells to release their stored fat for use as energy. Therefore, by producing less insulin, the body can concentrate on mobilizing and oxidizing fat stores as a fuel source for energy.

Considering the effect of carbohydrate on insulin and its attendant weight loss complications, a lot of diet promoters started taking advantage of this concept and applying it as the underlying principle behind the development of their diet plans.

The majority of these diets promote weight loss through the drastic reduction in the amount of consumed carbohydrates in order to substantially reduce insulin production.

In order to make up for the overall reduced energy occasioned by the reduced carbohydrate consumption, most of these diet plans promote the consumption of high amounts of either proteins or fats.

Although it is a fact that most of these high-protein and high-fat diets cannot be sustained for long, it would however be worthwhile embracing the underlying principle on which they are found - the reduction in insulin secretion.

Therefore, it is important to realize the fact that our food choices have a dramatic effect on our hormones which are the body's most powerful set of chemical messengers.

Consumption of the wrong types of foods is therefore bound to trigger undesirable hormonal variations such as the secretion of insulin which is often elevated through the consumption of high glycemic index carbohydrate foods.

In summary, the maintenance of a healthy and stable blood glucose level throughout the day should be of utmost importance to any individual who is seriously interested in losing weight and keeping it off in the long-term.

Effects of high insulin on the eye

Sugar is the primary source of energy to the cells and without this energy the cells starve. This is a major concern, especially to people with type 1 diabetes, and insulin replacement therapy is needed to facilitate this process.

High levels of sugar in the blood for many years has effects on many of the body's organs and systems. Thus many people with type 1 or type 2 diabetes will have a poor circulation system.

Their immune system may not be as responsive or efficient and they may have an increased risk of heart disease and atherosclerosis. Another common complication of diabetes is problems with the eyes.

The most common problem is known as diabetic retinopathy and is a consequence of the poor circulation. This chapter will discuss insulin and retinopathy and other eye problems that may occur.

Diabetes retinopathy is the most common complication of the disease in the eye. To a lesser extent a diabetic can suffer from cataracts and glaucoma but they are generally easy to detect and treat. Retinopathy is not that easy to spot and it is important that get a regular eye examination annually as part of the ongoing treatment for diabetes.

Diabetic retinopathy is categorized by the potential to lead to blindness.

Background retinopathy is the early stage of the condition. It is part of the micro vascular disease process whereby retinal blood vessels are affected by the thickening of the base membrane and the decrease in pericytes around the blood vessels.

This makes the blood vessels weak and they are prone to get micro aneurysms or small dilations in the blood vessel. They appear as red dots at the back of the eye. Over time they will disappear however they are a sign that the vessels are weakened and may burst or rupture at some later stage.

If this occurs blood is released to form retinal hemorrhages and hard exudates. Exudates are essentially scars from previous hemorrhages. If they occur in the macular area of the eye (where the lens focuses an image at the back of the eye) then vision loss will occur.

Proliferative retinopathy is the next stage. As there is less blood flow to the eyes the body compensates by producing more capillaries to supply blood to the eye. This is where the term proliferative comes from, the capillaries are multiplying across the eye, particularly in the vitreous body.

If these capillaries hemorrhage they will cloud the vision. As they heal they will clot or become fibrotic. They can pull the retina during this stage and cause retinal detachment and loss of vision.

There is no treatment for diabetic retinopathy, except for laser surgery to cauterize blood vessels and capillaries that may be a cause for concern. The other alternative is to keep blood sugar levels low and consistent.

As this is a never ending task it is important to understand how food affects blood sugar levels. You should implement a diabetic diet plan that includes eating regularly and eating the right kinds of foods. You should also implement an exercise regime as part of the ongoing treatment.

Effects of high insulin on the arteries

Heart and blood vessel diseases are amongst the worst known complications associated with Type 2 diabetes. People diagnosed with this form of diabetes are prone to high cholesterol and blood fat levels which can cause blood vessels to become clogged.

When the coronary arteries become clogged, the heart muscle itself can become starved for oxygen and various nutrients. When that happens, the heart can slow down or beat with less force, or both.

When the heart is unable to pump blood in sufficient quantities to feed the needs of the rest of the body, congestive heart failure develops. Blood that is unable to enter the heart because of congestion, backs up into the lungs which then become congested with fluid.

Exenatide, or Byetta, is a medication that mimics a kind of hormone known as an incretin. It stimulates the pancreas to release more insulin when the food just eaten starts to raise the blood sugar.

So, Byetta stimulates insulin secretion when blood sugars have actually risen after your meal, and the insulin secretion should stop when they drop. It also decreases the production of sugar by the liver.

Yes, it is often a complication, a clue from the effect of insulin resistance that shows your health care provider you have type 2 diabetes. It sneaks up on you; It often goes undetected for many years. Because of this, high blood sugar levels combined with elevated insulin levels may have already caused damage. Usually this damage is found in:

The blood vessels

Nerves

Eyes

Kidneys

The absolute impact on the health of the person with type 2 diabetes is through cardiovascular disease: the impact is two to fivefold. In addition type 2 is:

The major cause of blindness

Kidney failure

Amputations

Neurological complication such as impotency, and

Lifespan is decreased by an average of seven to twelve years

People with type 2 diabetes and heart disease differ from people with heart disease alone. Instead of high LDL's (bad) cholesterol, they usually have:

Moderately elevated LDL's

Low HDL's (good) cholesterol, and

High triglycerides levels

High blood pressure is also common and researchers suspect it is connected to insulin resistance. Hypertension increases the risk of heart attack or stroke because it puts more strain on the heart and blood vessels.

Undiagnosed type 2 diabetes is so widespread that researchers refer to the problem now as the "silent" epidemic or "silent" killer. Why is this you may ask?

It's because coronary artery disease is so far advanced before being identified it can no longer be aggressively treated due to years and years of silent heart damage. Diabetic patients with nerve damage often do not have chest pain, angina, which is the usual warning sign to get help before a heart attack occurs.

Complications can be avoided by recognizing the early signs of insulin resistance, being diagnosed and then maintaining blood sugar levels as close as possible to normal. A diagnosis of type 2 diabetes does not need to be a lifetime of suffering and premature death. This condition is treatable, and even reversible.

Effects of high insulin on the nerves and brain

Our main problem today is that virtually nobody consumes a small amount of fructose. Because sugar and high fructose corn syrup contain 50% fructose, most people consume large amounts of fructose from these sources as they have taken over our diet over the past forty years. Today many individuals get up to 40% of their calories from sugar or high fructose corn syrup.

When you consume this much fructose, the way it is metabolized in your body starts to change. Much of this excess fructose is converted to triglycerides and uric acid, two substances that can cause problems when you have too much of them in your body.

As you consume excessive fructose, your liver starts to fill up with fat from triglycerides and fat in the liver leads to insulin resistance where the cells in your body no longer respond appropriately to insulin when you consume a carbohydrate.

Insulin resistance is the first step down the pathway to type II diabetes. Excessive triglycerides also leads to central obesity where fat storage occurs in your abdominal cavity leading to the typical "beer belly".

High levels of uric acid irritate the linings of your blood vessels leading to something called endothelial dysfunction, the first step a process that can lead to heart attacks and strokes.

Uric acid also counters a gas called nitrous oxide that allows your blood vessels to relax. Thus if you have too much uric acid in your blood, the blood vessels contract rather than relax, leading to high blood pressure.

High blood pressure, heart attacks, strokes and beer belly obesity-not a pretty picture. But wait, it gets much worse. Remember the diseases listed above? How could they possibly tie into this metabolic mess? That's where it gets really interesting.

Once you have insulin resistance from consuming too much fructose, when you consume a carbohydrate, especially a rapidly absorbed one (high glycemic carbohydrate), you end up with a magnified glucose spike in your blood because the glucose can't easily enter into the cells in your body. And there's one organ in your body that definitely doesn't like magnified glucose spikes-your brain.

Unlike other organs in your body that can use substances other than glucose for energy, your brain is totally dependent on glucose for its energy needs.

Your brain is also unique in that it doesn't need insulin for glucose to enter into nerve cells.

Thus Mother Nature developed a complex system to keep the glucose levels in your blood at a steady level-too little glucose and your brain won't be able to function and too much glucose could damage sensitive nerve cells. That's because glucose is toxic to cells at high levels.

That's where magnified glucose spikes come into the picture. When you eat a Twinkie or other type of high glycemic carbohydrate, you end up with a magnified glucose spike in your blood and inside every nerve cell in your brain.

When this glucose is burned to produce energy, it releases too many free radicals-highly charged molecules capable of damaging the cell's delicate interior. Your natural built-in antioxidant system didn't evolve to handle this load of free radicals so you end up with nerve cells that don't work so well.

Magnified glucose spikes have another adverse effect on your brain cells-they cause the cells to dump out too many monoamine, neurotransmitters-dopamine, serotonin, norepinephrine and epinephrine so over time your brain cells don't have enough of these important chemicals to function properly.

You end up with brain cells that don't work very well leading to symptoms of brain dysfunction, the very symptoms that are typical for all the conditions listed above. There's another thing the brain doesn't do very well after it is exposed to magnified glucose spikes over a long period of time-regulate body composition.

If your brain is healthy, it works very hard to keep your total body fat stores in a narrow range. Although you can certainly override this auto regulatory function by continuously eating too much food when you aren't hungry, you would really have to work at it to get your body to store too much fat.

This became apparent to us when we directly measured the body composition of thousands of patients in our medical practice. We noticed that people with normal body composition almost never have the brain dysfunction symptoms indicating low levels of neurotransmitters.

When a person starts to develop these symptoms, within a short period of time the total amount of fat in their body slowly starts to increase. In other words their brain seems to flip them into a famine protective metabolic mode where they will store fat at any caloric intake.

Are you following me? Let's summarize this new disease concept. When you consume too much fructose over a long period of time, your liver starts to fill up with fat and you develop insulin resistance and abdominal obesity-the classic "beer belly".

Then when you consume rapidly absorbed or high glycemic carbohydrates, you end up with magnified glucose spikes in your blood and inside every nerve cell in your brain, eventually leading to diffuse brain dysfunction. As this brain disease progresses, you develop bothersome brain dysfunction symptoms and your body starts to store fat at virtually any caloric intake.

Most often we ignore headaches and try to suppress it with aspirin or paracetamol but do you know diabetes and headaches could be related? It could be an indication that you are approaching abnormal blood glucose level. Now think again, is it wise to ignore your headache? Or is it time to check out your glucose level?

When our body can't produce enough insulin to process the carbohydrates that we take with our everyday meal, glucose level rises. The carbs are actually turned into glucose in our body and the glucose is like fuel for our cells. We get energy because our cells keep functioning with this fuel.

Now, the problem is, glucose can't get into our cells by itself. Here comes the helper! Insulin!! Insulin takes this glucose into our cells. Without enough insulin secretion, we will end up having high glucose level in our blood; this is how we become diabetic!

And glucose as much as this, will lead to neuropathy or damage of nerves, exhaustion, severe headaches and if untreated this can cause death. So, don't you just avoid headaches from now on! Let's know about the types to stay alert.

Headaches during Hyperglycemia:

A patient is called hyperglycemic when too much sugar circulates in blood and as I have told earlier, the sugar level rises if there is not enough insulin to handle it. So, you have already figured out that hyperglycemia would be a hallmark of diabetes.

And sugar of this much could lead to damage of nerve fibers and blood vessels. Headache is the most common early symptom of hyperglycemia or diabetes and headaches of this type often come with blurred vision, exhaustion and sometimes confusion. So, don't ignore this one otherwise you have to pay a lot and suffer a lot.

Headaches during Hypoglycemia:

If hyperglycemia indicates high blood glucose level then obviously hypoglycemia will mean low blood glucose level. Hypoglycemia can even occur in diabetic patients too! Yes, no need to think that diabetic patients will always have their glucose level high! Usually this happens when the patient take more than the suggested dose of insulin or other diabetic medications.

A blunt headache with vision problems, dizziness, sweating, shudder and confusion are the symptoms of hypoglycemia and headache is the starter. Even healthy people could become hypoglycemic if they take lower amount of carbohydrates in their meal.

The quickest solution to hypoglycemia is to give the patient sugar or glucose as this can be easily metabolized by our body to maintain the glucose level. And remember, if not managed in the early stage, this hypoglycemia could result in convulsions; make you unconscious and could even lead to death.

Neuropathy headaches:

A high level of blood glucose could cause nerve damage or injury which is medically known as neuropathy. Neuropathy in diabetes patients could make the situation worse. Nerves which originate from our brain and brain stem are known as cranial nerves and these cranial nerves could be the victim of this neuropathy incidence especially in diabetes patients which will ultimately cause severe hammering headaches.

Most shockingly, doctors don't apprehend these headaches often or misdiagnose as migraines and as a result both diabetes and headaches remain untreated until something really bad happens. Here are some tips for immediate measurement when the headaches start.

Massage the forehead starting from the centre and keep doing it until the pain eases or have a cup of coffee. We all know caffeine works like magic by decreasing the swelling of blood vessel, thus relieving pain. And whenever there is a headache, a cold press on forehead always helps.

Well, the best solution is to not get a headache by keeping the glucose level under control! Not so high and not so low. You need to avoid taking extra or very low carbs and be careful about taking insulin shots and, last but not the least don't forget to check on your glucose level.

INSULIN AND HORMONE relationships

Many people do not realize the connection between hormones and between hormones and diet. There are definite links between having balanced hormones and diet. One of the key links is the connection between insulin and cortisol.

Many people have gotten into some very bad habits that caused their diets to go off kilter. Back in the 1990s, people developed a fear of ingesting fat. As a result, they turned to low-fat foods, in addition to things such as refined carbohydrates and simple sugars.

This trend caused a major problem; and, it actually created a situation in which their bodies had to use a lot of insulin to balance all the sugar they were putting into their bloodstreams.

It eventually got to the point where their bodies weren't listening to the insulin and in fact became insulin resistant. This is an ongoing problem. The effect: The Obesity Epidemic

Dr. Alicia Stanton, MD functional medicine physician, and author of Hormone Harmony. She shed light on the issues related to insulin and cortisol.

The issue with insulin being out of balance is twofold. One problem is the direct relationship between insulin and belly fat. The more insulin you have, the more belly fat you tend to put on. This is a medical issue because belly fat has different enzymes in it that unbalance the hormones.

Belly fat takes testosterone and converts it to extra estrogen, creating a big problem for both men and women. Women with too much extra estrogen can have breast tenderness, heavy bleeding, uterine fibroids, and other problems. That's why it's important to keep estrogen and progesterone levels balanced.

An imbalance between estrogen and progesterone, especially estrogen dominance, can occur at any age. It is tragic to see children develop insulin resistance or Type 2 diabetes, which is traditionally known as adult-onset diabetes. In addition, these children probably have a genetic predisposition to diabetes, in addition to a poor diet.

The other issue with respect to hormone balance is the fact that cortisol is the balancing hormone for insulin. This means that if insulin is out of whack, cortisol also gets out of whack.

A very real hormone connection: Cortisol can affect your hormones, and progesterone, the building block for cortisol, also happens to be the building block for estrogen and testosterone.

What does this mean? If you have a big demand for cortisol due to your body's attempts to balance insulin, and you take all your building blocks and push them over towards cortisol, you won't have the building blocks necessary to make estrogen and testosterone. You also won't be able to balance your estrogen with the progesterone.

In a nutshell, this is the simple connection between the diet and the hormones. It's important to watch your diet so that the hormones won't become unbalanced and cause serious health problems. Of course, it's all very complicated and specialists in functional and metabolic medicine can do the proper testing to help you get back in balance and on the right track.

Estrogen and insulin have a synergistic relationship, so it's virtually impossible for a woman to balance her hormones if her blood sugar is not also balanced. But when she does balance her blood sugar, then her other hormones often become balanced on their own.

Insulin is an important hormone that is released by the pancreas, and its job is to make the energy from sugar in your blood available to your cells in the form of glucose. The balancing act is a very delicate one, and many factors can interfere with it.

In fact, eating too much of the wrong processed foods can tax the pancreas so much that it stops producing insulin properly, and this is a serious disease called diabetes.

There are three key things you can do to help balance your blood sugar:

Eat?

While this may seem obvious, many women have a deeply-ingrained adversarial relationship with food.

This often shows up as either overweight or underweight, but even women of normal weight can have a concern (if not an obsession) with their weight, and habitually skip meals, snack on now-quality food, or just under eat. Or many women are so busy taking care of others that they just don't stop to feed themselves.

Without getting too technical here, just know that processed, simple carbohydrate foods have a high glycemic index (GI), and these cause your blood sugar to spike, which tells the pancreas to produce lots of insulin, which then causes your blood sugar to sharply drop. Not only does this feel lousy, but all that insulin interferes with your estrogen production and also contributes to weight gain.

In fact, many "low fat" food and snacks that women eat because they think those are "healthy" may in fact be causing their blood sugar and insulin to jump around, having quite the opposite of a healthy effect.

On the other hand, foods with a low GI (in general, foods containing protein and healthy fats) promote a much slower, more stable release of blood sugar and insulin. These are actually much better choices for both hormone balance and weight management.

There are problem proteins in some foods called lectins, and these have harmful effects on the body. Lectins cause things to stick together, and they literally make the blood stickier, increasing the risks for stroke and other health concerns. They also contribute to inflammation, impair immunity, promote weight gain, and mimic insulin.

For each blood type, there are a handful of foods that contain these harmful lectins. In my experience, when people stop eating the foods with lectins for their blood types, everything tends to improve. And specifically, not only do women tend to lose bloat and body fat, but their hormone balance dramatically improves.

Nutrients and insulin balance

Overwhelming scientific evidence confirms that vitamin, mineral and antioxidant deficiencies suppress immune function and contribute to chronic inflammatory degenerative processes, such as arthritis, cancer, Alzheimer's, cardiovascular disease and diabetes.

We have seen over and over the myriad of scientific studies that demonstrate the ability to prevent, treat and even reverse type 2 diabetes through diet and lifestyle.

Lifestyle intervention has also been demonstrated to be more effective than metformin (the most commonly prescribed medication for this condition) for reducing the incidence of metabolic syndrome, pre-diabetes and type 2 diabetes.

Unfortunately, however, serious nutritional deficiencies can occur even with a healthy eating plan due to many possible factors, including inadequate absorption of nutrients due to pre-existing medical conditions, toxins in our environment that may interfere with transport of nutrients, or nutrient-drug interactions as a result of the use of both prescription and over-the-counter medications.

Metformin, for example, has been shown to lower vitamin B12 and folate, which can then lead to an increase in homocysteine levels, a major risk factor for cardiovascular disease.

A folate deficiency has also been associated with the diabetes complications of retinopathy (a leading cause of blindness) and renal failure (kidney diseases). Many other pharmaceutical medications and over-the-counter treatments have been implicated in causing micronutrient deficiencies.

Although, fresh, whole (unprocessed) foods are the basis of this lifestyle intervention approach to treat and reverse chronic disease, we cannot ignore that a number of supplemental nutrients have been shown to improve blood sugar control and reduce insulin resistance, and other inflammatory processes, such as alpha-lipoic acid, chromium, vitamin D, riboflavin and niacin. This is not an exclusive list, however, as there are many more micronutrients that can be considered to effect glycemic management.

Warnings:

Always keep in mind that supplementation of nutritional supplements without proper diagnostic testing and monitoring is absolutely contraindicated. Routine conventional serum concentration measurements are not what is meant by proper diagnostic testing. Most micronutrient deficiencies go undetected with conventional serum testing until a severely malnourished state exists.

Instead, specialized micronutrient laboratory testing that assess the white blood cells is necessary since it will evaluate not just the levels of a nutrient, but also the function of a nutrient that is present in blood or tissue. This type of testing can be obtained through specialized laboratories, and is not always reimbursed by health insurance, but has more recently become more affordable.

The investment in this type of testing can prevent unnecessary and costly supplementation use, as well as possibly prevent the need for prescription medications since it can pinpoint specific individual requirements and get at the root of the problem.

Micronutrient testing needs to be performed before supplementation is considered and repeat testing should occur at least every six months when supplementation is indicated. Dosing will vary individually and may not necessarily be in line with the Recommended Dietary Allowances (RDA) for that nutrient.

Repeat diagnostic testing is necessary so that you can assess if you are getting the correct dose. Repeat testing is also necessary in order to assess if you are affecting other micronutrients, as micronutrient supplementation cannot be in isolation.

As an example, with too much chromium supplementation, nutrient-nutrient interactions can occur, such as decreased zinc absorption, which can then worsen glycemic control. Vitamin C can cause increased chromium absorption, and iron metabolism can be affected since chromium competes with iron for transport.

Nutrients in excess can also have serious effects, such as too much chromium can cause kidney and liver toxicity, rhabdomyolyisis (breakdown of muscle tissue), psychiatric disturbances, and hypoglycemia.

Although safety is seldom an issue for nutritional supplements, just like pharmaceutical medications, nutritional supplements can also carry risks and side effects, especially when taken with other pharmaceutical agents.

Remember to also test your blood glucose often when taking supplements that are indicated to lower blood glucose, in order to reduce the risk of hypoglycemia, especially if taking along with pharmaceutical medications or other herbs and supplements that may affect blood glucose levels.

Nutritional supplements must be used cautiously and with your healthcare providers knowledge. This is not "cookbook" medicine. Your health needs are different from your neighbors needs therefore always.

Sensible Supplementation:

When using any nutritional supplementation, you must take a sensible approach to avoid risks and side effects:

• Start low and go slow but give it a honest try for at least one month

• Try one new supplement at a time (every 3-4 weeks)

• Only take what you need (not every potential supplement)

• Test your blood sugar often to monitor effects

• Don't take at the same time as your other medications

• Contact your health care professional for adjustment in your other medications if needed, due to low blood sugar and/or medication interactions

• Purchase only quality products and avoid supplements with artificial sweeteners, artificial colors, and binders (that often include genetically modified organisms [GMOs])

Chapter 7

Cholesterol

MANY PEOPLE ARE AWARE that high levels of cholesterol are dangerous to their health because of the increase risk of heart disease and stroke. But what is cholesterol? And is it really bad for the health?

What is cholesterol? It is a combination of fatty substances and steroid naturally produced by the body. In fact, cholesterol is in every cell of our body. It makes cell membranes less fluid and holds it together without the need for a cell wall. It basically keeps the cell in the body stable.

Cholesterol is also a source of energy. The liver produces 80 percent of the cholesterol but it can also be from dietary source. It also helps in digesting fats and in vitamins absorption.

What is cholesterol doing in our body? There are good and bad cholesterol. Bad cholesterol of low density lipoprotein (LDL) is produced by our body and is also needed, but high level of bad cholesterol as mentioned can cause heart disease and clog the arteries. LDL are usually found in animal meat, ice cream, eggs and butter so limit the intake of these food.

What is cholesterol which known as HDL? You may heard about this before, High density lipoprotein (HDL) is also the good one which protects you from heart disease by lowering the bad cholesterol in the body.

Foods high in HDL are sea foods like salmon and tuna. Foods high in fiber like fruits and vegetables also helps raise the HDL level in the body. What is cholesterol doing in our body is really important, eating more vegetables and high fiber food and limiting red meat and sweets can definitely help lower your bad LDL.

Cholesterol is not really bad for the body; it only becomes a threat if increases beyond the normal level. The truth is what is cholesterol is an essential substance which our body needs to work properly. We won't be able to function without cholesterol.

The most important is the need to maintain a proper diet and exercise to make sure the good cholesterol is in a healthy level. You should also visit your doctor regularly so that you can monitor your cholesterol level and if you might need medication or a change in your diet for your cholesterol.

What is cholesterol is basically what you take in your body. If you eat unhealthy food then you'll get bad cholesterol if you eat a balanced diet coupled with exercise you'll have higher good cholesterol. It's a good lifestyle that will ensure a good health and a good level of cholesterol.

It is extremely important for each and every single individual to learn the basics regarding cholesterol levels. Everyone stands the chance of developing high cholesterol. Here, we will take a glance at cholesterol and how it can affect you.

Cholesterol is created naturally in the body by the liver. However, there are many different types of foods that we can acquire cholesterol from as well. These foods include things like eggs, various types of diary products, and even different types of meat.

Many individuals are under the impression that cholesterol is a bad thing. This is not true. While there is a "good" cholesterol type, and a "bad" cholesterol type, this substance is needed for the proper functioning of the body.

The "bad" type of cholesterol is often referred to as "LDL". This stands for low-density lipoprotein. While this is an essential component to the body, excess levels of it can result in serious health complications, such as heart disease. Extremely low levels of LDL cholesterol have been found to result in different types of cancers.

The "good" type of cholesterol is often referred to as "HDL". This stands for high-density lipoprotein. It has been determined that this type of cholesterol is beneficial in protecting an individual against heart problems, and other health risks. This is due to the fact that it helps to eliminate the amount of "bad" cholesterol in the body and pushes LDL through the arteries.

Cholesterol is an essential in the process of producing hormones within the body. In addition to this, cholesterol also helps to create a steady supply of vitamin D in the body and increases the amount of acids that help to break down bile. This is important to the body as it helps to process and digest fat that the body is exposed to.

If an individual has too much cholesterol that is considered "bad" in the body, many different things occur. The first thing that happens is that the body starts to accumulate a buildup of plaque in the arteries. As time progresses, this results in what many refer to as the "hardening" of the arteries, or "atherosclerosis".

When the arteries experience this type of complication, the passageways may become narrowed. This narrowing of the passageways may become so severe that the artery becomes obstructed.

This will decrease the mobility of the blood in the body. As a result, the heart and other important organs of the body will fail to receive the proper amount of oxygen and nutrients.

When the body does not receive the necessary oxygen, nutrients, and other substances in the body, then that can result in the failure of the functionality of these essential body components. Many who have experienced this issue have experienced a heart attack, stroke, and other serious medical complications.

In addition to LDL and HDL cholesterol, there are two other types of cholesterol. The next type is called "VLDL". This is also called "very low density lipoproteins". This type is quite high when it comes to the amount of fats that are contained in it, but extremely low when it comes to protein.

The fourth type of cholesterol is called "triglycerides". Specifically, this particular item is actually a "fat". When an individual experiences VLDL, triglycerides are carried through the blood. This can pose a number of health risks for an individual.

There are many different factors that can affect the cholesterol levels of an individual. The first factor is an individual's weight level. People who are overweight, or considered to be "obese" normally have high cholesterol levels. It is important to maintain a weight that is appropriate to the body mass index in order to avoid this complication.

The next thing that can affect a person's cholesterol levels is their family medical history. It has been established that high cholesterol can actually run in the family. If an individual has a genetic predisposition of high cholesterol levels, it is likely that they will experience it as well.

If an individual is gaining in years, the possibility of acquiring high cholesterol is a possibility. It has been determined that as an individual ages, the possibility of experiencing high cholesterol is high. If a female experiences menopause, their chances of developing high cholesterol are also elevated.

People who have low levels of physical activity are prone to developing high cholesterol levels. It has been established that the average person should engage in approximately three to four hours of exercise on a weekly basic in order to increase the amount of HDL cholesterol in the body.

The next factor that can greatly increase the amount of LDL, or "bad" cholesterol in the body is an individual's diet. Diet has one of the largest impacts on cholesterol levels. People who consume a healthy diet of various types of fatty fish, a variety of nuts, and soy-based products are less likely to have high cholesterol complications.

There are a number of medical conditions that can result in high levels of cholesterol. For example, many people who experience the devastating condition of diabetes may experience high levels of LDL cholesterol. It is important for these individuals to appropriately manage their sugar levels to decrease this possibility.

Many individuals pay no attention to their cholesterol levels. This can be a very unfortunate experience for many. For every five people in the world, there is at least one person who experiences high amounts of LDL cholesterol, and low amounts of HDL cholesterol. This issue should be taken very seriously.

Individual's who suffer from the effects of high cholesterol levels can develop cardiovascular disease. People who suffer from this medical complication can experience a serious heart attack, and/or a stroke.

While many may live to speak of their experience with a heart attack and/ or stroke, there are many who do not. Cholesterol can actually turn fatal for many people around the world!

If you are concerned about your health, you should make an appointment with your doctor to determine if you have high cholesterol. If you find that you do, you should work closely with the treatment guidelines that your doctor outlines for you.

LDL Cholesterol Levels

There are various factors which affect your cholesterol levels. It's either your age, sex, hereditary or food intake. Whatever your reason might be, keeping your cholesterol level under control is very essential.

You may have heard your physician asking you to reduce the "bad cholesterol" from your body. LDL cholesterol levels are supposed to be the "bad" ones amongst the other cholesterol moving in your blood stream.

Why you are usually asked to lower your "bad cholesterol" levels?

It might make you wonder how you can find a solution to lower your cholesterol levels. What exactly is this "bad cholesterol"? Does "good cholesterol" exist, too? What is the difference between the two cholesterols? These are the various questions which disturb your mind constantly.

That's because, your health and maybe even your life depends on it. Cholesterol is a form of lipid or fat which circulates along with your blood stream. A cholesterol unit constitutes two different elements, the lipids and the proteins, hence they are usually called lipoproteins.

There are three major types of lipoproteins moving along with your blood flow. LDL or low density lipoprotein, HDL or high density lipoprotein and VLDL or very low density lipoprotein.

LDL and VLDL are categorized as "bad cholesterols" while HDL is categorized as "good cholesterol". This categorization is due to their bodily functions or capabilities. Incredible HDL benefits the body while excess of LDL and VLDL harms your body drastically.

How to control your LDL cholesterol levels to control bodily damage?

Cholesterol is produced naturally by your liver to provide multiple benefits but excess of it through your dietary intake might cause extensive damage to your body. Cholesterol helps in producing vitamins and hormones, structuring of your body cells and providing energy to your body.

As HDL performs the function of carrying cholesterol away from the arteries and back to the liver, it is tagged as "good cholesterol". LDL is tagged as "bad cholesterol" because it tends to gather the excess around the walls of the arteries.

LDL accumulation tends to form cholesterol plaques around the walls of the arteries, damaging the arterial walls and obstructing the natural blood flow. This gives rise to severe consequences such as extreme strain on heart muscles, elevated blood pressure, thickening and narrowing of the arteries and increased risk of cardio vascular diseases.

When cholesterol plaques are torn it forms blood clots which obstructs the smooth blood flow. Due to no blood supply to the heart, you might experience a heart attack and no blood supply to the brain can cause a stroke.

High LDL cholesterol levels and cardio vascular diseases is more likely for a man over the age of 45 years and for a woman over the age of 55 years or who is experiencing menopause. For a child, the cholesterol levels can increase due to overweight or hereditary issues.

Foods That Caused High Cholesterol Are Common Foods We Eat Everyday

What are the foods that caused high cholesterol? You are probably someone who thought your health was great. You exercise and you eat proper portions. After taking a blood test at your doctor's office, you were told that you have high cholesterol.

What may have caused such to happen? It is an issue you must address quickly, for it may turn your health upside down. The thing you must do is understand what you are eating. It is not about the amount of food you take in, but what kind of food you eat.

It is advisable you keep a food journal with you; write down everything you eat over a period of time. This is great to do so you can see what food you eat and make the necessary changes and adjustments to your diet. Knowledge is the key. When you know what foods cause your cholesterol to be high, you can avoid them and make more educated healthy choices along the way.

One of the first things to realize is that anything that says it is "cholesterol free," is not. These products may not have cholesterol in them, but they have trans fats which your body will eventually turn into bad cholesterol. Keep that in mind. Margarine is high in that type of fat that the body changes to cholesterol. Mayonnaise is the same thing. It makes sandwiches taste better, but it is all trans fats. Whole milk or cheeses cause high cholesterol. Basically any dairy product causes high cholesterol.

Red meat most of all contain the most cholesterol. It also happens to be something we eat and enjoy. This is probably the main culprit in causing your high cholesterol. Switch to white or lean meats. Fish will be the best replacement for it.

These are examples of foods that caused high cholesterol. It is not just about food with high cholesterol in them, but foods that contain the fats which will prompt your body to change it into cholesterol. Now that you know what to avoid, compare them to your food journal and make the changes needed. Strive to eat healthy and you will live a healthier life.

Choosing the Right Food to Reduce Cholesterol

We are very busy persons: We have things to do, problems to solve, people to talk to and it seems that we stopped having enough time to eat. More and more often full course meals get replaced for quicker snacks.

Snacking does not necessarily represent a bad thing. Snacks are normal for a body that consumes its inner energy for its daily activities and needs to accumulate more strength.

Eating wrong foods for snacks is causing you trouble like increasing cholesterol level. It is the snacks high in carbohydrates and fats that are causing you an elevated cholesterol and weight gain. This situation will eventually lead to heart disease, heart attacks, and diabetes.

In order to reduce cholesterol level, you must pay attention to what you eat and how you eat. Here are a few basic rules that you have to follow if you need to reduce cholesterol, basically if you want to have a healthy diet.

1. Cut back on foods rich in unsaturated fat like and fried foods. Try eating more often lean meat and fish.

2. If you want to reduce cholesterol level, you need to eat more vegetables. There are never too many vegetables in an elimination process. These foods present the advantage of having a low calories level, but, above all, they contain a broad number of vitamins and antioxidants that are helping your body fight cellular damage.

3. You don't need to give up eating dressings or gravies. You just have to reduce their quantity in your elimination. Therefore you should start serving them aside to make sure you control the amount you're eating. You can even buy low fat dips.

4. Pay attention to how you cook your food. Stop using partially hydrogenated vegetable oils (found in margarines and shortenings) as they contain trans-fatty acids, a form of fat that is raising cholesterol level. Cookies, candies, and chips also contain these trans-fatty acids, so be careful. Grilling your food instead of frying it will surely help you reduce cholesterol.

5. Carbohydrates give you quick energy as they are converted into sugar in your body, but they are also causing you weight gain and high glucose level that will eventually lead to diabetes. They also lower your HDL level. So be careful of carbohydrates. Whole grain products are low in flour and high in fiber, minerals, and vitamins so eating them will reduce cholesterol and make you healthier.

6. If you are badly craving for a snack, have some nuts and seeds. They contain unsaturated fats that are lowering the LDL level (bad cholesterol) which eventually reduce total cholesterol level. Nuts and seeds also give you vitamin E, B (which will help you have a healthy shiny hair and strong nails), and minerals.

These rules may make you think like there is no fun in eating anymore. Keep in mind that you don't have to give up on 'wrong' foods for good. You just have to eat healthy and with moderation. And once in a while you are allowed to indulge yourself with a moderate amount of your favorite 'not-so-healthy' snack.

Chapter 8

___

Glycemix Index

THE GLYCEMIC INDEX (GI) is a numerical system of measuring how fast a carbohydrate triggers a rise in circulating blood sugar-the higher the number, the greater the blood sugar response. A low GI food will cause a small rise, while a high GI food will trigger a dramatic spike.

The glycemic index (GI) is a ranking of carbohydrates on a scale from 0 to 100 according to the extent to which they raise blood sugar levels after eating.

Foods with a high GI are those which are rapidly digested and absorbed and result in marked fluctuations in blood sugar levels. Low-GI foods, by virtue of their slow digestion and absorption, produce gradual rises in blood sugar and insulin levels, and have proven benefits for health.

Low GI diets have been shown to improve both glucose and lipid levels in people with diabetes (type 1 and type 2). They have benefits for weight control because they help control appetite and delay hunger. Low GI diets also reduce insulin levels and insulin resistance.

The glycemic index (GI) is a measure of the power of foods (or specifically the carbohydrate in a food) to raise blood sugar (glucose) levels after being eaten.

The GI values of foods must be measured using valid scientific methods. It cannot be guessed by looking at the composition of the food. Currently, only a few nutrition research groups around the world provide a legitimate testing service.

The GI value of a food is determined by feeding 10 or more healthy people a portion of the food containing 50 grams of digestible (available) carbohydrate and then measuring the effect on their blood glucose levels over the next two hours.

For each person, the area under their two-hour blood glucose response (glucose AUC) for this food is then measured. On another occasion, the same 10 people consume an equal-carbohydrate portion of glucose sugar (the reference food) and their two-hour blood glucose response is also measured.

A GI value for the test food is then calculated for each person by dividing their glucose AUC for the test food by their glucose AUC for the reference food. The final GI value for the test food is the average GI value for the 10 people.

Foods with a high GI score contain rapidly digested carbohydrate, which produces a large rapid rise and fall in the level of blood glucose. In contrast, foods with a low GI score contain slowly digested carbohydrate, which produces a gradual, relatively low rise in the level of blood glucose.

Glycemic Index is an important indicator of the healthiness quotient of foods. It is especially helpful for diabetic patients as foods rich in carbohydrates are known to have a significant impact on the diabetic level of the patient.

The most common misunderstanding most people have with the glycemic index, is that they think they're supposed to only eat certain things and avoid other things. Most diets and eating plans work this way. The glycemic index however, is designed to help you make better eating choices, not tell you what to eat or not eat.

The good glycemic foods; that is, those with the lower rates, are more desirable not only for diabetics, but for those who are watching their carbohydrate intake through such diets as the South Beach Diet, they should also be aware of what the good glycemic foods are.

The Glycemic Index ranks foods on how they affect the blood glucose levels. It measures the amount of increase in your blood glucose levels two to three hours after eating.

The Glycemic Index shows how quickly a single food will turn into blood glucose on a scale of 100. Pure glucose is given a value of 100.

As you know, the Glycemix Index ranks how a single food breaks down in your body and is converted to blood glucose after eating. But if you follow the rules of performance nutrition, you should be eating a 'complete' meal:

That consists of a lean protein

A starchy carbohydrate

And a fibrous carbohydrate

Not a single food by itself.

This changes the Glycemix Index of that meal!

Second...

When you consume proteins with carbohydrates, it can greatly lower the blood glucose effects of that food. A baked potato's score 85 on the Glycemic Index when combined with a protein is much lower.

Third...

There are flaws of the Glycemic Index like:

Limited data. Only about 5% of the foods are listed in the Glycemic Index. And there is a very limited number of researchers that currently do testing.

The numbers on the Glycemix Index are an average of the responses of groups of people. This explains the variation in some charts. The numbers listed are not exact values.

A wide variation of in the actual Glycemic Index measurements. For example, a baked Russet potatoes have been tested with a Glycemic Index as low as 56 and as high as 111!

Food preparation methods like microwaving, grinding, frying, baking, etc. There's even differences in the GI when boiling pasta for 10 minutes or 15 minutes.

Food combinations can affect the Glycemic Index of a listed food. While the Glycemic Index is based on single foods, we often consume foods in combinations. This can affect the overall Glycemic Index of that meal. Figuring out the precise Glycemic Index of foods after being mixed is less accurate.

Individual differences in a response to a food on the Glycemic Index. People simply have different blood glucose responses. Without monitoring each person's actual blood glucose levels, results can and often will vary.

Reliance on the Glycemic Index can lead to over eating. If you only rely on the Glycemic Index to pick and choose foods you can end up consuming too many fats and excess calories.

In any event, the Glycemic Index is useful to people with certain dietary needs. But its service to the bodybuilder is vague. There's no way that refined pasta is better for you than natures own potato.

Basing your choices only on the Glycemic Index can lead to over consumption of high calorie foods. And with the limited data and varied testing results, your own reactions to a particular food may vary greatly.

The Glycemic Index - Why It's So Important For Weight Loss

For many years, people were told to either cut down on their calories or eat low fat food. In recent years though, the discovery of the Glycemix Index has shed new light on the truth about how to lose weight.

The problem with cutting down on calories, is that if you suddenly eat less calories than your body will react to this by slowing down your metabolism (the rate at which you burn up energy). So this means that although some people experience an initial drop in weight, they will struggle to keep that weight off long term.

The other thing to take into consideration with cutting down on calories, is that it doesn't take into account how healthy the source of those calories is. For example, I could eat just one chocolate bar a day and eat far fewer calories than someone eating loads of fruit and vegetables all day, but I think it's obvious which of us would be healthier.

The problem with low fat diets is that our bodies need fat, but the right kind of fat. Saturated, monounsaturated fat, polyunsaturate fat, and omegas 3, are good fats that are needed for optimum health.

So what does this Glycemic Index thing mean then? Well, put simply, the glycemic index measures how fast a food is converted into sugar for use as energy by our body.

If our body gets a rush of too much sugar all at once, the pancreas releases insulin to take the extra sugar out of the blood, where it is then stored temporarily as Glycogen in your muscles and liver. However, this is only a temporary store, to be used as immediate energy. If you do not do enough physical activity to use up this energy, it eventually gets converted into fat.

By choosing foods that have a low glycemic index, you can instead get a steady release of sugar into your bloodstream. This then reduces the likelihood of there being too much sugar in your blood at any time, and therefore there is less chance of any excess sugar eventually being stored as fat.

Chapter 9

Sweeteners

———

THERE ARE NOW DOZENS of sweeteners on the market or listed in the ingredients of food and drink; and with each one itself potentially having many different names, it can be a minefield to unravel the definitions and understand the implications for our health. There are some important categories and definitions to understand if we are to successfully negotiate this minefield.

Sugars vs. Sweeteners

Sugars are carbohydrates that have particular molecular characteristics - there are many varieties and they have widely differing properties (In fact, strictly speaking all carbohydrates are sugars, but we do not need to worry about that here.)

Sweetener is a more general term, referring to any substance that can be used to make something taste sweeter. This includes some sugars - for example, glucose. Other sweeteners, such as Aspartame, are not sugars.

Sugar the Nickname vs. Sugar the Category

This can be confusing for newcomers to the topic and is the most important to get straight from the start. As mentioned above, there are many kinds of sugar. However, when people talk about 'Sugar', especially in the context of food, they are often referring to Sucrose, one particular sugar. Certainly when you see the word in the list of ingredients for a product, this is what it means. Sucrose has acquired the nickname 'Sugar' over the years because it is the most commonly used sugar. It has acquired many other names too, but this is the most common.

A good illustration of this is a dried fruit like figs. You will not see the word Sugar in the ingredients list of a packet of dried figs - yet in the nutritional breakdown it might say carbohydrate, of which sugars - 65g. This is because whilst there is no sucrose in the figs, there is naturally occurring fructose and glucose - which are themselves sugars.

Artificial vs Natural

The term natural sweeteners is typically understood to mean substances that can already be found in plants or animals, unlike artificial sweeteners, which cannot. However, natural sweeteners often only occur in very small amounts in nature and undergo much processing before they find their way into our foods.

For example, fructose is a natural sweetener - it is responsible for some or all of the sweetness in most fruits. Yet it is also found in high fructose corn syrup, a highly processed sweetener derived from corn. Likewise, Tagatose is found in small amounts in dairy products, but used as a sweetener in concentrated form.

Acesulfame potassium (also known as Acesulfame K, Sunett, Sweet One or E950) is an artificial sweetener, often found in canned drinks marketed as being sugar free. It was formulated by a German chemical company and is not found in plants or animals.

Nutritive vs Non-nutritive Sweeteners

The definition of nutritive sweeteners is that they have calories, whereas non-nutritive sweeteners do not. This is broad distinction and not always clear cut, so it should not be taken as a guide to whether a sweetener is the right choice, because within these categories there are wide variations.

Nutritive Sweeteners

There are two families of nutritive sweeteners - sugars and sugar alcohols.

Sugars - this family of sweeteners is commonly found in (and extracted from) naturally sweet foods; as such they are also classed as natural sugars. Their names usually end with -ose; for example, Sucrose, Glucose and Fructose.

Glucose is found in many fruits and along with fructose is responsible for their sweet taste. Sugars tend to be the highest calorie sweeteners and some of them have been implicated in causing or exacerbating health problems such as tooth decay, heart disease, obesity and diabetes.

Sugar Alcohols - this family of sweeteners have names that end in -ol; for example, Sorbitol and Xylitol. Sugar alcohols are classed as natural sugars - Xylitol, for example, is found in some fruits and vegetables as well as in the bark of the Birch tree.

Most sugar alcohols have fewer calories than sugars because they only partially digest. This can have intestinal implications and common side-effects of over consumption include diarrhoea and flatulence. However, their low impact on blood sugar, reduced calories and the fact they do not cause tooth decay makes them a widely-used substitute for sugars.

An example of how these categories can blur is the sugar alcohol Erythritol. It does not get absorbed by the gut at all and therefore has no calories - yet as a sugar alcohol it is classed as a nutritive sweetener.

Non-Nutritive Sweeteners

For the most part, non-nutritive sweeteners are artificial sweeteners. Examples are Aspartame (Phenylalinine), Saccharin and Sucralose. The reasons why a non-nutritive sweetener might not provide calories include:

1. It does not digest, but passes through unchanged - for example Saccharin

2. It digests, but has no calorific value - for example sucralose

3. It does digest and does have calories, but is so sweet that the amounts required to sweeten a product are tiny - for example Neotmame, which is 10,000 times sweeter than sucrose.

Non-nutritive sweeteners are typically found in products marketed as sugar free. In some cases (for example, Aspartame) there have been studies that suggest large quantities can be harmful when fed to rats.

One non-nutritive sweetener that is not artificial is Rebaudioside A (Stevia). Stevia is a non-nutritive sweetener because it is not metabolised by the body, but it is also a natural sweetener, since it can be found in the South American stevia plant.

Different Kinds of Sweetener Alternatives

Throughout history, humans have been adding natural sweeteners to food in order to make it more palatable. Up until today, people still use natural sweeteners as one of the primary ingredients in food. In nature, there are different kinds of natural sweeteners that you can use to make every meal you eat delicious in taste.

However, you also need to consider the fact that science has found a way to make synthetic sweetener or artificial sweetener. During the past decades, artificial sweeteners proved to be much cheaper than natural sweeteners.

In fact, every household in western countries uses some kind of artificial sweeteners as food additive. Because of the development of artificial sweeteners, the natural ones have been neglected and are now less used in many households in western countries today.

We are also growing more and more concern about their health and the kind of foods we eat. Sweet foods are now considered to be unhealthy and can cause different kinds of illnesses if taken in a long term basis.

However, what people don't know is that this is mainly because of the effect of artificial sweeteners that is now widely common in western households and the civilized world.

Research has found that it is not actually the sweet food that causes different kinds of diseases but it is mainly caused by the additives in artificial or synthetic sweeteners. Further research suggests that natural sweeteners are healthier than artificial ones.

However, some people say that artificial sweeteners bring out more of the flavor on food. What they don't realize is that natural sweeteners are as good as the artificial ones regarding the taste and also the potency.

There were also findings that natural sweeteners have less calories and are considered to be healthier than the artificial ones. Natural sweeteners are recommended for diabetics and it is also recommended for people who are on a diet but still loves to eat sweets.

So, here are the different kinds of natural sweeteners that you can add to your meals to make it more delicious without worrying about the bad effects on your body:

Stevia - This is a type of sweetener used by South Americans and because it contains fewer calories than other common sweeteners, it is now growing very popular in the United States. This particular sweetener can be added to almost any of your food and beverages that needs to be sweetened. This particular sweetener comes in powder tablets or liquid

Tagatose - This is another type of natural sweetener that is found in milk. Tagatose has the same chemical composition as fructose but in terms of chemical and physical characteristics, it is quite different.

Agave - This particular sweetener is very popular among health and diet fanatics. It has fewer calories than artificial sweeteners and is recommended for diabetics. This particular sweetener can act as a substitute to honey. However, Agave is not as thick and as sweet as honey but many people have said that it has a wonderful taste that you will truly love.

These are some of the different kinds of natural sweeteners available in the market today. So, if you want a healthier alternative to artificial or synthetic sweeteners, you should consider getting some of the natural sweeteners mentioned. It will provide you with better tasting food and also effectively managing diabetes and minimize the risk of rotting teeth.

What you need to know about food sweeteners - Natural or artificial sweeteners side effects

With all the sweetening options available today, reliable, natural table sugar has been pushed to the back of the cupboard in favor of artificial sweeteners, raw sugar or honey and molasses for adding sweetness.

"When it comes to sugar and cane sugar and any other word they're using, sugar is sugar." Good to know as you're planning those sweet (and delicious) holiday recipes.

The most well-known artificial sweeteners - saccharin, aspartame and sucralose - are sold under the names Sweet 'N Low, Equal, NutraSweet and Splenda, as well as being part of sugar free foods and diet drinks.

While these sweeteners do deliver sweetness without the calories, some scientific studies using animals have raised questions about the link between aspartame, saccharine and cancer. That's hard to ignore, and if you're concerned, you might want to limit your intake of foods and beverages containing these ingredients.

It is very important to distinguish between naturally occurring sugar and processed sugar, as the effects of the two can be quite different. Naturally occurring sugar is not something you sprinkle onto your food, it is already in the food.

Naturally occurring sugar can be found in grains, beans, vegetables and fruits and because it has not been processed, still contains a variety of vitamins, minerals, enzymes and proteins.

When properly cooked, chewed and digested, the natural sugars in foods are broken down and enter the bloodstream to provide energy. However, once sugar goes through processing (which is the case with refined table sugar) it is stripped of all vitamins, minerals and fiber.

As a result, extra effort is required to digest the sugar, and the body must use its own stores in order to break it down, leaving the body depleted of essential nutrients. Rather than providing nutrition, sugar instead causes deficiencies.

It is because of this speed at which sugars can be digested, and the resulting depletion of nutrients that leads to the infamous sugar high, followed by the notorious sugar crash.

In truth, sugar can affect ones health in many different ways. As seen in the list below, many are not direct results of the sugar itself, but rather a result of our addiction to sugar. For example, if you drink soda, you will not necessarily become obese right away.

In fact, you may not become obese in a year...or even two! However, the sugar in the soda will decrease your body's ability to recognize when it is full, leading you to eat larger and larger meals.

In addition, the sugar in the soda will train your body to crave super sweet foods, leaving you feeling unsatisfied with healthier alternatives. The end result is a lifetime of weight gain and decreased health.

When the public became very aware of calories in their food, and were looking to consume as little calories as possible, they noticed that sugar had a significant amount of calories. In order to continue to manufacture and consume sweet foods, zero-calorie products were created to replace sugar.

These products come in the form of artificial sugar substitutes, also known as artificial sweeteners. Since artificial sweeteners are not a naturally occurring food, they must be approved before being added to foods as per the Food Additives.

The FDA has currently approved of five artificial sweeteners:

Aspartame, sold under the brand names NutraSweet and Equal.

Saccharin, sold under the brand name Sweet'N Low.

Sucralose, sold under the brand name Splenda.

Acesulfame K (or acesulfame potassium), produced by Hoechst, a German chemical company; widely used in foods, beverages and pharmaceutical products around the world.

Neotame, produced by the NutraSweet Company; the most recent addition to FDA's list of approved artificial sweeteners, neotame is used in diet soft drinks and low-calorie foods.

Sweeteners out there that are not as damaging to the body and blood sugar levels. Just to clarify, they are still a type of sugar and have to go through some form of processing.

Therefore, even natural sweeteners should be consumed in moderation. Yet, since many still have fiber, vitamins and minerals intact, they are a better choice than any processed or artificial sweetener.

Here is a list of more natural sweeteners. If you have never tried, or even heard of, any of these, expand your taste buds and give some a shot. You will be shocked at how delicious they are and how easily you can make the switch!

Agave Nectar

Barley Malt

Birch Sugar

Birch Syrup

Brown Rice Syrup

Date Sugar

Honey

Maple Syrup

Maple Sugar

Molasses

Rapadura

Stevia

Sucanat

Vegetable Glycerin

Now that you are armed with knowledge, one very important thing to remember is that the body craves balance. Completely stripping your life of sugar and all situations where you may come face to face with sugar can end up being more stressful, and therefore damaging, then creating a healthy relationship with sugar.

Even though the studies on artificial sweeteners do not concretely link them to dangerous symptoms, they are still a chemically derived substance. Humans are not meant to consume chemicals. The human body is very complex and to find a connection between an artificial sweetener and a symptom is very difficult to prove.

There are so many other factors that can come into play, such as stress, overall diet, and physical activity level, that those conducting the studies cannot say that the artificial sweetener is the sole cause. However, if you simply look at what these products are made of and all the possible side effects, that is enough of a reason to avoid them at all costs.

Processed sugar may not pose the serious threats that artificial substitutes do, but it may still come with its own side-effects. Avoid processed sugar as much as possible, but try to avoid isolating yourself in doing so. Lastly, naturally occurring sugars are your best choice.

Not only will they have the least side effects, but they also will save you money in the long run: Less doctors' visits, and lower grocery bills because you simply will not eat as much food when you start incorporating real, whole foods into your diet.

Sweet foods that contain naturally occurring sugar will help to lessen your overall sugar cravings. These foods also contain vitamins, minerals, enzymes and fiber that help the body in the assimilation of sugar. However, even here the rule of "everything in moderation" still applies. Too much of a good thing is no longer a good thing!

Overall, think of your need for sugar as a craving for sweetness in your life. Your body craves balance and needs love, humor and rest just as much as stability, focus and activity. If you live a stressful, busy life, make a point to add more sweetness (maybe in the form of hugs and kisses) to your life, and you may end up not needing any sugar at all!

PART TWO: MIMICKING DIET MEAL PLAN

Chapter 10

Fast-Mimicking Days

─────

FIVE CONSECUTIVE DAYS a month, you will eliminate animal protein and limit calories to 900 a day to mimic the benefits of a whole month of full-time calorie restriction.

Fast-Mimicking Days You now know how important calorie restriction is for your health and longevity. The great news is that you can restrict calories for only five consecutive days out of the month and still reap the benefits of an entire month of calorie restriction. That's right. Valter Longo, head of the Davis School of Gerontology at the University of Southern California, has shown that a monthly five-day modified vegan fast gives you the same longevity-boosting results as a month of a traditional calorie-restricted diet does. I strongly recommend that you begin the Longevity Diet program by doing five fast mimicking days in a row. Not only will you get the same benefits as if you had restricted your calorie intake for the whole month, but you will dramatically change the makeup of your gut bacteria in those five days, driving out the bad bugs and nourishing your gut buddies. In fact, research by Dr. Longo and by researchers at the University of Colorado, Boulder, confirms that a fast-mimicking diet changes the types of bacteria in your gut for the better. In a study conducted by researchers at the Knight Lab at the University of Colorado, participants who followed just a three-day cleanse saw dramatic shifts in their microbiome, including increased levels of our good friend Akkermansia! And once your gut buddies are in good shape, they will make it much easier for you to follow the rest of the program. Please don't worry that you will starve to death by cutting calories for a few days. Humans can easily go without food for two months and longer if water is available. If you do find yourself getting hungry and really needing something more, feel free to have about a tablespoon of MCT oil up to three times a day to ward off the "low-carb flu." As you grow more insulin sensitive and your gut buddies take over and start renovating, you will be able to complete these five days easily without hunger becoming an issue. So what will you be eating during these five days? Actually, the things you won't be eating are far more important, so let's start there.

Foods to avoid

All dairy products All grains and pseudograins All fruit, including all seeded vegetables, which are technically fruits All sugar sources Unapproved seeds Eggs Soy products Nightshade plants (eggplant, peppers, tomatoes, potatoes) Corn, soy, canola, and other vegetable oils Meat, chicken, and all other animal products.

Foods to include

And what can you eat? Vegetables

You can eat as much as you'd like of all the following vegetables, either cooked or raw. If you have irritable bowel syndrome, SIBO, diarrhea, or another gut issue, limit your consumption of raw veggies and cook the rest of the things you eat thoroughly. All vegetables should be organic and can be purchased either fresh or frozen. If fresh, they should be in season and grown locally with sustainable farming practices, if at all possible. Cruciferous vegetables: Bok choy, broccoli, Brussels sprouts, Swiss chard, any color and type of cabbage, cauliflower, kale, mustard greens, collard greens, rapini, kohlrabi, watercress, mizuna, arugula. Greens of all kinds: Belgian endive, all kinds of lettuce, spinach, dandelion greens, chicory Treviso, radicchio Artichokes Asparagus Celery Fennel Radishes and other root vegetables such as yams, taro root, jicama, yucca, cassava, turnips, rutabagas, horseradish Fresh herbs: Mint, parsley, sage, basil, and cilantro, plus garlic and all kinds of onions, including leeks and chives Ocean vegetables: Kelp and seaweed, including sheets of nori.

Protein

For these five days, you are going to go vegan. That means no eggs, meat, chicken, or dairy products of any kind. Do not worry that you will become protein deficient! Remember, you are probably eating too much protein right now, and your body recycles the protein that is already present.

Eliminating animal products for five days gives your body a rest from digesting all that protein and allows it to become an eco-friendly! Sources of plant based protein that you can eat during these five days include but do not have to include: Tempeh (fermented soy, without grains) Hemp tofu and hemp seeds, legumes such as lentils and beans, nuts and seeds.

Fats and oils

Acceptable vegetable fat sources for these five days include: Avocado—feel free to have a whole one each day. First-cold-pressed extra-virgin olive oil. Olives of any kind.

Nuts: Walnuts, macadamia nuts, pistachios, hazelnuts, pine nuts, almonds, blanched almond flour. Avocado oil, coconut oil, macadamia nut oil, MCT oil perilla oil, sesame seed oil, walnut oil, hemp seed oil, flaxseed oil.

Condiments and seasonings

Because of their sugar content (not to mention other harmful ingredients), avoid all commercially prepared salad dressings and sauces. Instead, use as much as you like of the following. Fresh lemon juice, vinegars, mustard freshly ground black pepper, iodized sea salt. Your favorite herbs and spices.

Beverages

Obviously, you will avoid all sodas (including diet soda), sports drinks, lemonade, and other commercially prepared beverages. Instead, enjoy at least eight glass of tap or filtered water a day, as well as: San Pellegrino or other Italian sparkling mineral water (or Acqua Panna, a still mineral water) As much tea as you'd like—green, black, or herbal. Regular and/or decaffeinated coffee (black or with unsweetened almond, hemp, or coconut milk). Stevia extract. Just Like Sugar (inulin), or monk fruit to sweeten your tea or coffee, if you like Using the following meal plan, you will duplicate the effects of a month of calorie restriction while stimulating stem cell regeneration and strengthening your gut wall.

Chapter 11

———

Mimicking Diet Menu

DAY 1

BREAKFAST Green Smoothie

SNACK Romaine Lettuce Boats Filled with Guacamole

LUNCH Arugula Salad with Hemp Tofu, Grain-Free Tempeh, or Cauliflower "Steak" and Lemon Vinaigrette

SNACK Romaine Lettuce Boats Filled with Guacamole

DINNER Cabbage-Kale Sauté with Grain-Free Tempeh and Avocado

Day 2

BREAKFAST Green Smoothie

SNACK Romaine Lettuce Boats Filled with Guacamole

LUNCH Romaine Salad with Avocado, Coriander Pesto, and Grain-Free Tempeh

SNACK Romaine Lettuce Boats Filled with Guacamole

DINNER Lemony Brussels Sprouts, Kale, and Onions with Cabbage "Steak"

Day 3

BREAKFAST Green Smoothie

SNACK Romaine Lettuce Boats Filled with Guacamole

LUNCH Hemp Tofu-Arugula-Avocado Seaweed Wrap with Coriander Dipping Sauce

SNACK Romaine Lettuce Boats Filled with Guacamole

DINNER Roasted Broccoli with Cauliflower "Rice" and Sautéed Onions

Day 4

BREAKFAST Green Smoothie

SNACK Romaine Lettuce Boats Filled with Guacamole

LUNCH Leek Soup

SNACK Romaine Lettuce Boats Filled with Guacamole

DINNER Hemp Tofu-Arugula-Avocado Seaweed Wrap with Coriander Dipping Sauce

Day 5

BREAKFAST Green Smoothie

SNACK Romaine Lettuce Boats Filled with Guacamole

LUNCH Creamy Cauliflower Parmesan Soup

SNACK Romaine Lettuce Boats Filled with Guacamole

DINNER Cauliflower "Fried Rice"

Chapter 12

Mimicking Diet Recipes

GREEN SMOOTHIE

This is the perfect breakfast during your five-day "fast" and on free days as well. Add a little more water if the smoothie is too thick. You can make a triple batch and refrigerate for up to three days in a covered glass container.

Serves 1

1 cup chopped romaine lettuce

1/2 cup baby spinach

1 to 3 fresh mint sprigs, with stems

1/2 avocado

4 tablespoons freshly squeezed lemon juice

3 to 6 drops liquid stevia, to taste

1/4 cup ice cubes

1 cup tap or filtered water

Place all the ingredients in a high-powered blender and blend on high until smooth and fluffy, adding more ice cubes if desired

---

ROMAINE LETTUCE BOATS Filled with Guacamole

I recommend that you use Hass avocados for your guacamole (and other recipes). Hass have a black or dark green pebbly skin and contain more fat (the healthy monounsaturated kind).

Serves 1

SERGIO GUZZARDI

1/2 avocado

1 tablespoon finely chopped red onion

1 teaspoon finely chopped fresh coriander

1 tablespoon freshly squeezed lemon juice

Pinch of sea salt, preferably iodized

4 large romaine lettuce leaves, washed and patted dry

Place the avocado, onion, coriander, lemon juice, and salt in a bowl and mash with a fork until smooth. To serve, scoop an equal amount of the guacamole into each lettuce leaf.

Arugula Salad with Hemp Tofu, Grain-Free Tempeh, or Cauliflower "Steak" and Lemon Vinaigrette

This is another great option for your five-day "fast" that you can put together easily to take to work for lunch or throw together for dinner at the end of a long day.

Serves 1

For the tempeh:

1 tablespoon avocado oil

4 grain-free tempeh, cut into 1/2-inch-thick strips

1 tablespoon freshly squeezed lemon juice

1/4 teaspoon sea salt, preferably iodized

For the dressing:

2 tablespoons extra-virgin olive oil

1 tablespoon freshly squeezed lemon juice

Pinch of sea salt, preferably iodized

For the salad:

1 and 1/2 cups arugula

Zest of 1/2 lemon (optional)

MAKE THE TEMPEH:

In a small skillet, heat the avocado oil over medium heat. Place the tempeh strips in the hot pan and sprinkle them with the lemon juice and salt. Sauté the tempeh strips for about 2 minutes; turn them and sauté for another 2 minutes, until cooked through. Remove from the pan and reserve.

MAKE THE DRESSING:

Combine all the dressing ingredients in a mason jar with a tight-fitting lid (double the ingredients if making two batches). Shake until well combined.

MAKE THE SALAD:

Toss the arugula in the dressing and top with the tofu, tempeh, or cauliflower steak, adding the lemon zest, if desired.

OTHER VEGETARIAN VERSIONS:

In place of the tempeh, tofu, or cauliflower, substitute acceptable Quorn products: Chik'n Tenders, Grounds, Turk'y Roast, Chik'n Cutlets. (They contain a tiny amount of egg white so are not totally animal protein free, but the amount is probably negligible in terms of mTOR.)

Cabbage-Kale Sauté with Grain-Free Tempeh and Avocado

This tasty dish makes a great substitute for a grain bowl and is very adaptable. Be sure to use bok choy or Napa cabbage instead of green cabbage if eating this during your five-day "fast."

Serves 1

1/2 avocado, diced

3 tablespoons freshly squeezed lemon juice

4 pinches of sea salt, preferably iodized

3 tablespoons avocado oil

1 and 1/2 cups thinly sliced green cabbage

1/2 red onion, thinly sliced

4 ounces grain-free tempeh

Toss the avocado in 1 tablespoon of the lemon juice and season with a pinch of salt. Set aside. Heat a skillet over medium heat. When it is hot, add 2 tablespoons of the avocado oil and the cabbage and onion. Sauté until tender, about 10 minutes, stirring occasionally. Season with 2 more pinches of salt. Using a slotted spatula, remove from the skillet and set aside. Add the remaining 1 tablespoon avocado oil to the skillet, raise the heat to high, and add the remaining 2 tablespoons lemon juice and the tempeh. Sear the tempeh, flipping after 3 minutes, until cooked through, about 6 minutes total. Season with the remaining pinch of salt. To serve, top the sautéed cabbage and onions with the tempeh and avocado.

OTHER VEGETARIAN VERSIONS:

In place of the tempeh, tofu, or cauliflower, substitute acceptable Quorn products: Chik'n Tenders, Grounds, Turk'y Roast, Chik'n Cutlets. (They contain a tiny amount of egg white so are not totally animal protein free, but the amount is probably negligible in terms of mTOR.)

OTHER VEGAN OPTIONS:

Replace the grain-free tempeh with hemp tofu or a cauliflower "steak" (a 3/4-inch-thick cauliflower slice seared over high heat in avocado oil until golden brown on both sides)

Romaine Salad with Avocado, Coriander Pesto, and Grain-Free Tempeh

This satisfying salad will keep you full and energized on the five-day "fast." To save time, make the coriander pesto in advance and store for up to three days in the refrigerator in a covered glass container. You can substitute basil or parsley for the coriander.

Serves 1

For the tempeh:

1 tablespoon avocado oil

4 grain-free tempeh, cut into 1/2-inch-thick strips

1 tablespoon freshly squeezed lemon juice

1/4 teaspoon sea salt, preferably iodized

For the pesto:

2 cups chopped coriander

1/4 cup extra-virgin olive oil

2 tablespoons freshly squeezed lemon juice

1/4 teaspoon sea salt, preferably iodized

For the dressing:

1/2 avocado, diced

2 tablespoons freshly squeezed lemon juice

2 tablespoons first-pressed extra-virgin olive oil

Pinch of sea salt, preferably iodized

SERGIO GUZZARDI

For the salad:

1 and 1/2 cups chopped romaine lettuce

MAKE THE TEMPEH:

In a small skillet, heat the avocado oil over medium heat. Place the tempeh strips in the hot pan and sprinkle them with the lemon juice and salt. Sauté the tempeh strips for about 2 minutes; turn them and sauté for another 2 minutes, until cooked through. Remove from the pan and reserve.

MAKE THE PESTO: Place all the pesto ingredients in a high-powered blender. Process on high until very smooth.

MAKE THE DRESSING:

Toss the avocado in 1 tablespoon of the lemon juice and set aside. Combine the remaining 1 tablespoon lemon juice, the olive oil, and the salt in a mason jar with a tight-fitting lid. (Double the ingredients if making two batches.) Shake until well combined.

MAKE THE SALAD:

Toss the romaine in the dressing. Arrange the avocado and tempeh over the lettuce and spread the pesto on top.

Lemony Brussels Sprouts, Kale, and Onions with Cabbage "Steak"

You can use any of the many types of kale in this hearty vegetable dish. Unless you're using baby kale, remove the stems before chopping. (There is no need to remove the stems of or chop baby kale.)

Serves 1

4 tablespoons avocado oil

One 1-inch-thick red cabbage slice

1/4 teaspoon plus 1 pinch of sea salt, preferably iodized

1/2 red onion, thinly sliced

1 cup Brussels sprouts, thinly sliced

1 and 1/2 cups chopped kale

1 tablespoon freshly squeezed lemon juice

First-pressed extra-virgin olive oil (optional)

Heat a skillet over high heat. When it is hot, add 1 tablespoon of the avocado oil, reduce the heat to medium, and sear the cabbage slice until it is golden brown on one side, about 3 minutes. Flip it and brown it on the other side. Season with the pinch of salt, remove to a plate, and cover to keep warm. Wipe the skillet clean with a paper towel and return it to the stovetop. Heat 2 tablespoons of the avocado oil in the skillet over medium heat. Add the onion and Brussels sprouts. Sauté until tender, about 3 minutes. Add the remaining 1 tablespoon avocado oil, the kale, and the lemon juice and sauté for another 3 minutes, until the kale is wilted. Season with the 1/4 teaspoon salt. To serve, top the cabbage "steak" with the sautéed vegetables. Add a drizzle of olive oil, if desired.

Hemp Tofu-Arugula-Avocado Seaweed Wrap with Coriander Dipping Sauce

Nori is a form of seaweed that has been flattened into squares or strips. It makes a terrific stand-in for flatbread in this wrap, which can be eaten as part of your five-day "fast."

Serves 1

For the filling:

1 tablespoon avocado oil

4 ounces hemp tofu, cut into 1/2-inch-thick strips

2 tablespoons freshly squeezed lemon juice

1/4 teaspoon sea salt, preferably iodized, plus additional to taste

1/2 avocado, diced

For the dipping sauce:

2 cups chopped fresh coriander

1/4 cup first-pressed extra-virgin olive oil

2 tablespoons freshly squeezed lemon juice

1/4 teaspoon sea salt, preferably iodized

For the wraps:

1 cup arugula

1 sheet nori (sushi seaweed)

4 green olives, pitted and halved

Sea salt, to taste

MAKE THE FILLING:

In a small skillet, heat the avocado oil over medium-high heat. Place the hemp tofu strips in the hot pan and sprinkle with 1 tablespoon of the lemon juice and the salt. Sauté the strips for about 2 minutes; turn them and sauté for another 2 minutes, until cooked through. Remove from the pan and reserve. Toss the avocado in the remaining tablespoon lemon juice and season with salt. Set aside.

MAKE THE DIPPING SAUCE:

Place all the dipping sauce ingredients in a high-powered blender. Process on high until very smooth.

MAKE THE WRAPS:

Spread the arugula over the bottom half of the seaweed sheet. Top with the filling and olives. Sprinkle with salt to taste. Carefully roll into a tight wrap, sealing the end with a little water. Cut in half and serve with the coriander dipping sauce.

OTHER VEGETARIAN VERSIONS:

In place of the tempeh, tofu, or cauliflower, substitute acceptable Quorn products: Chik'n Tenders, Grounds, Turk'y Roast, Chik'n Cutlets. (They contain a tiny amount of egg white so are not totally animal protein free, but the amount is probably negligible in terms of mTOR.)

OTHER VEGAN VERSIONS:

Replace the hemp tofu with grain-free tempeh or a cauliflower "steak" (a 3/4-inch-thick cauliflower slice seared over high heat in avocado oil until golden brown on both sides).

A bamboo mat, available in the Asian foods section of most supermarkets, can help you roll tight seaweed wraps

Roasted Broccoli with Cauliflower "Rice" and Sautéed Onions

To make cauliflower "rice," grate the cauliflower with a cheese grater, using the largest holes, into rice-shaped pieces. You can also pulse it in a food processor, using the S blade and being careful not to overprocess it. If you use a food processor, cut the cauliflower into chunks first.

Serves 1

1 and 1/2 cups broccoli florets

2 and 1/2 tablespoons avocado oil

3 pinches of sea salt, preferably iodized

1/2 head medium cauliflower, riced

1 tablespoon freshly squeezed lemon juice

1/4 teaspoon curry powder

1/2 red onion, thinly sliced

Preheat the oven to 180° C. Put the broccoli into a Pyrex dish with 1 tablespoon of the avocado oil. Roast in the oven for 15 minutes, stirring frequently, until tender. Season with a pinch of salt.

Sauté the cauliflower in a medium skillet with 1 tablespoon of the avocado oil, the lemon juice, the curry powder, and a pinch of salt until tender, 3 to 5 minutes. Do not let it get mushy by overcooking. Transfer the cauliflower "rice" to a plate and wipe the skillet clean with a paper towel. Reheat the skillet over medium heat. When it is hot, add the remaining 1/2 tablespoon avocado oil and the sliced onion and sauté until tender, stirring frequently, for about 5 minutes. Season with a pinch of salt. To serve, place the cauliflower "rice" on a plate and top with the broccoli and sautéed onions.

Longevity Leek Soup

The leeks in this soup are a wonderful longevity food with loads of polyphenols. It has a bright lemony flavor with a richness from the nutmeg that will keep you warm all day.

Serves 4 to 6

2 tablespoons extra-virgin olive oil

1 pound leeks, cleaned and chopped

2 stalks celery, diced

3 cloves garlic, minced

1 tablespoon chopped fresh thyme

Zest of 1 lemon

1 large head cauliflower, cut into 2-inch florets

1/2 teaspoon fresh nutmeg

1 teaspoon fine sea salt, or more to taste

2 teaspoons coarse black pepper

2 quarts vegetable stock

1 bay leaf

Finely chopped chives for garnish

In a large soup pot, heat the olive oil over medium-high heat. Add the leeks, celery, garlic, thyme, lemon zest, and cauliflower along with the nutmeg, salt, and pepper, and sauté over medium heat, stirring regularly until the leeks begin to wilt. Add the stock and the bay leaf and cook, covered, for 25 to 35 minutes, until the cauliflower is very tender. Blend using a stick blender, or transfer into a regular blender and blend until smooth (work in batches so as not to overfill the blender). Once pureed, return to the heat and cook for an additional 10 to 15 minutes. Taste, and adjust seasoning as needed. Serve garnished with chopped chives.

Creamy Cauliflower Parmesan Soup

If you love a leek and potato soup or chowder, chances are this soup will be right up your alley. Plus, it's full of cruciferous cauliflower and brainboosting olive oil.

Serves 6

3 tablespoons extra-virgin olive oil

1 sweet onion, minced

2 stalks celery, diced

3 cloves garlic, minced

2 large heads cauliflower, cut into 2-inch florets

1/2 teaspoon fresh ground nutmeg

1 teaspoon fine sea salt, or to taste

2 teaspoons coarse black pepper

1 tablespoon white miso paste

7 cups mushroom

Stock or

2 cups coconut milk

1/4 cup grated Parmesan cheese or nutritional yeast

1 bay leaf

Finely chopped chives or thyme for garnish

In a large soup pot, heat the olive oil over medium-high heat. Add the onion, celery, garlic, and cauliflower, along with the nutmeg, salt, and pepper, and sauté over medium heat, stirring regularly, until the leeks begin to wilt. Add the miso paste and cook, stirring, until the paste is incorporated. Add the stock, coconut milk, Parmesan, and bay leaf and cook, covered, for 35 to 45 minutes, until the cauliflower is very tender. Blend using a stick blender, or transfer into a regular blender and blend until smooth (work in batches so as not to overfill the blender).

Once pureed, return to the heat and cook an additional 10 to 15 minutes. If it is too thick, thin with a little water. Taste and adjust the seasoning as needed. Serve garnished with chopped herbs and additional grated Parmesan.

Cauliflower "Fried Rice"

This is a great, filling meal to enjoy during your five-day "fast." With options like this, you'll never go hungry even while your body thinks you're fasting.

Serves 6 to 8

2 tablespoons sesame oil

1 medium brown onion, diced

1/4 cup minced spring onions

1-inch piece ginger root, peeled and minced

2 cloves garlic, minced

1 cup thinly sliced mushrooms (any type)

4 ribs celery, thinly sliced

1 cup broccoli florets

4 ounces water chestnuts (canned okay), roughly chopped

4 cups cauliflower rice

1 tablespoon coconut aminos

1/4 teaspoon paprika

1/4 teaspoon powdered mustard

In a large skillet or wok, heat the oil over medium-high heat. Add the onion, spring onions, and ginger and cook several minutes, until the onions are translucent. Add the garlic, mushrooms, celery, broccoli, and water chestnuts and cook, stirring frequently, until the vegetables soften and the garlic is fragrant (5 to 6 minutes). Turn heat to high and add the cauliflower rice. Cook for 3 to 4 more minutes, stirring frequently to ensure that it doesn't burn.

After a minute, add the coconut aminos, paprika, and powdered mustard. Continue cooking on high heat, stirring frequently, until the cauliflower is tender but not mushy, and serve.

Free days

After you finish your five-day "fast," you can begin the free days portion of the plan and eat as much as you'd like. If you go from the five-day "fast" back to your old habits, particularly sugar consumption, bad bugs can regrow quickly, undoing much of the progress you made. On free days, you do not have to limit calories, but you should be mindful of your protein intake. As Dr. Valter Longo say that you need only 0.37 gram of protein per kilogram of body weight. Therefore, you can easily meet your protein needs for an entire day with a scoop of whey protein powder, a couple of eggs, one protein bar, or about 80 grams of pastured chicken or wild fish. To make it simple, on your free days, I want you to focus on eating, at maximum, one roughly 80 grams serving of protein. You choose whether that means eggs at breakfast, a salad with tuna at lunch, or a small serving of wild fish or shellfish with dinner. Please start to view grassfed and -finished beef as an occasional treat rather than as a diet mainstay. At your other meals, you'll get plenty of protein from veggies, nuts, mushrooms, and lentils and by recycling the mucus in your gut. Other than that, feel free to enjoy any and all of the following foods.

Longevity-promoting acceptable foods

Oils

Olive oil

Algae oil

Coconut oil

Macadamia oil

MCT oil

Avocado oil

Perilla oil

Walnut oil

Red palm oil

Rice bran oil

Sesame oil

Flavored cod-liver oil

Sweeteners

Stevia

Just Like Sugar (made from chicory root [inulin]) Inulin Yacon

Monk fruit

Luo han guo (aka monk fruit)

Erythritol

Xylitol

Nuts and Seeds (1/2 cup per day)

Macadamia nuts

Pili nuts

Baruka nuts

Walnuts

Pistachios

Pecans

Coconut (not coconut water)

Coconut milk (unsweetened dairy substitute)

Coconut milk or cream (unsweetened, full-fat canned)

Hazelnuts

Chestnuts

Brazil nuts (in limited amounts)

Pine nuts

Flaxseeds

Hemp seeds

Hemp protein powder

Psyllium seeds or powder

Olives All

Coconut Yogurt (plain)

Dark Chocolate 72% cacao or greater (30 grams per day)

Vinegars All

Herbs and Seasonings All

Miso Bars

Flours

Coconut

Almond

Hazelnut

Sesame (and seeds)

Chestnut

Cassava

Green banana

Sweet potato

Tiger nut

Grape seed

Arrowroot

Ice cream

Coconut milk dairy-free frozen dessert

Pasta

Cappello's gluten-free fettuccine, lasagne and gnocchi

Shirataki noodles and rice

Wine red (180 ml per day)

Spirits (30 ml per day)

Dark spirits like bourbon, scotch, dark tequila, dark rum, cognac, gin. Avoid vodka.

Fruits (limit all to their seasons except avocado)

Avocados

Blueberries

Raspberries

Blackberries

Strawberries

Cherries

Crispy pears (Anjou, Bosc, Comice)

Pomegranates

Kiwis

Apples

Citrus fruits (no juices)

Nectarines

Peaches

Plums

Apricots

Figs

Dates

Vegetables

Cruciferous Vegetables:

Broccoli

Brussels sprouts

Cauliflower

Bok choy

Napa cabbage

Chinese cabbage

Swiss chard

Arugula

Watercress

Collards

Kohlrabi

Kale Green and red cabbage

Raw sauerkraut

Kimchi

Other vegetables

Treviso radicchio

Chicory

Curly endive

Nopales cactus leaves

Celery

Onions

Leeks

Chives

Scallions

Carrots (raw)

Carrot greens

Artichokes

Beets (raw)

Radishes

Daikon radish

Jerusalem artichokes (sunchokes)

Hearts of palm

Coriander

Parsley

Okra

Asparagus

Garlic

Mushrooms

Leafy greens

Romaine

Red- and green-leaf lettuce

Mesclun (baby greens)

Spinach

Endive

Dandelion greens

Butter lettuce

Fennel

Escarole

Mustard greens

Mizuna

Parsley

Basil

Mint

Purslane

Perilla

Algae

Seaweed

Sea vegetables

Resistant starches

Tortillas (only those made with cassava and coconut flour or almond flour)

Bread and bagels made by Barely Bread

Paleo Wraps (made with coconut flour)

The Real Coconut Café Tortillas and Chips

In Moderation

Green plantains

Green bananas

Baobab fruit

Cassava (tapioca)

Sweet potatoes or yams

Blue or purple sweet potatoes

Rutabaga

Parsnips

Yucca

Celery root (celeriac)

Glucomannan (konjac root)

Persimmon

Jicama

Taro root

Turnips

Tiger nuts

Green mango

Millet

Sorghum "popcorn"

Green papaya

Plant-based "meats"

Quorn: Chik'n Tenders, Grounds, Chik'n Cutlets, Turk'y Roast, Bacon-Style

Tempeh (grain free only)

Legumes

Lentils (preferred)

Black soybeans

Chickpeas

Adzuki beans

Other beans

Peas

Disease-Promoting, Life-Shortening Foods to Avoid

Refined, starchy foods

Pasta

Potatoes

Potato chips

Milk Bread

Tortillas

Pastry

Wheat, rye, barley, rice, quinoa, soy, corn flours Crackers

Cookies

Cereal

Sugar

Agave

Splenda (sucralose)

NutraSweet (aspartame)

Sweet'N Low (saccharin)

Diet drinks

Maltodextrin

Vegetables

Peas

Sugar snap peas

Legumes

Green beans

Chickpeas (including as hummus)

SOY PRODUCTS

Tofu

Edamame

Soy protein

Nuts and seeds

Pumpkin seeds

Sunflower seeds

Chia seeds

Peanuts

Cashews

Fruits (some are incorrectly called vegetables)

Cucumbers

Zucchini

Pumpkins

Squash (any kind)

Melon (any kind)

Eggplant

Tomatoes

Bell peppers

Chili peppers

Goji berries

NON–SOUTHERN EUROPEAN Cow Milk Products (these contain casein A1)

Yogurt (including Greek yogurt)

Ice cream

Frozen yogurt

Cheese

Ricotta

Cottage cheese

Kefir grains, sprouted grains, pseudograins, and grasses

Wheat (pressure-cooking does not remove lectins from any form of wheat)

Einkorn wheat

Farro

Kamut

Oats (cannot pressure-cook away the lectins)

Quinoa

Rye (cannot pressure-cook away the lectins)

Bulgur

White rice

Brown rice

Wild rice

Barley

Buckwheat

Kashi

Spelt

Corn

Corn products

Corn syrup

Popcorn

Wheatgrass

Barley grass

Oils

Soy

Grape seed

Corn

Peanut

Cottonseed

Safflower

Sunflower

Canola

Acceptable Animal Protein Sources in Limited Amounts

Dairy Products (30 grams cheese or 120 ml yogurt per day)

Real Parmesan (Parmigiano-Reggiano)

French or Italian butter

Buffalo butter

Ghee

Goat yogurt (plain)

Goat milk as creamer

Goat cheese, goat butter

Goat or sheep kefir

Sheep cheese and yogurt (plain)

Aged French, Italian, or Swiss cheese

Buffalo mozzarella

Casein A2 milk (as creamer only)

Organic heavy cream

Organic sour cream

Organic cream cheese

Fish (wild caught; 4 ounces per day maximum)

Whitefish, including cod, sea bass, redfish, red or pink snapper

Freshwater perch, pike

Alaskan halibut

Canned tuna

Alaskan salmon

Hawaiian fish, like mahi-mahi, opakapaka

Shellfish (wild caught)

Shrimp

Crab

Lobster

Scallops

Calamari (squid)

Clams

Oysters

Mussels (farmed okay)

Abalone (farmed okay)

Sea urchin (uni)

Sardines

Anchovies

PASTURED POULTRY (NOT free range; 120 grams per day)

Chicken

Turkey

Goose

Duck

Pheasant

Quail

Ostrich

Pastured, non-soy or -corn fed, or omega-3 eggs (up to 4 daily), but limit whites, e.g., make an omelet with 4 yolks and 1 white)

Meat (grass fed and finished; 4 ounces per day maximum; once per week maximum)

Bison

Wild game

Venison

Boar

Elk

Pork (humanely raised or pastured)

Lamb

Beef

Prosciutto

Bresaola

Liver and other organ meats

PART TREE: KETOGENIC DIET RECIPES

Chapter 13

BAKED GOODS

1. BAKING MIX

Ingredients

- 385g almond flour

- 455g plan whey protein powder

- 55g flaxseed meal

- 70g baking powder

- 20g salt

- 110g refined coconut oil

---

28 SERVINGS – Per serving:

198 calories; 10g fat; 21g protein; 9g carbohydrate; 2g dietary fiber; 7g net carbohydrate

Directions

1. Put all the dry ingredients in your food processor and pulse until everything is evenly distributed.

2. Add the coconut oil.

3. Pulse until its cut in; you'll want to scrape down the sides of the processor a couple times.

4. Store in a snap-top container in a cool place.

2. Muffin

Ingredients

• 450g Baking Mix (above)

• 120ml unsweetened coconut milk

---

12 SERVINGS – Per serving:

132 calories; 7g fat; 14g protein; 6g carbohydrate; 1g dietary fiber; 5g net carbohydrate

Directions

1. Preheat the oven to 220°C. Coat a muffin tin well with cooking spray.

2. In a bowl, whisk the Bake Mix and coconut milk together.

3. Divide the batter among the muffin cups-I used my cookie scoop, and it was just the perfect size, one scoop per biscuit. My scoop holds 2 tablespoons (28g).

4. Bake for 8 minutes.

5. Serve hot with plenty of butter.

3. Waffles

Ingredients

- 460g Baking Mix

- 315ml unsweetened coconut milk

- 1 egg

---

12 WAFFLES – Per serving:

143 calories; 8g fat; 15g protein; 6g carbohydrate; 1g dietary fiber; 5g net carbohydrate

Directions

1. Combine the bake mix and coconut milk in a mixing bowl, preferably one with a pouring lip, and whisk until the lump are gone.

2. Let the batter sit for 5 minutes, during which time it will thicken a bit.

3. While the batter is thickening, put your large, heavy pan over medium heat. You want it hot before you add the batter.

4. The biggest pan will fit three 10-cm pancakes. Melt 2 tablespoons (28g) of butter in the pan and then poured in the batter.

5. Serve them with more butter and perhaps a sprinkle of erythritol and cinnamon. Or serve with Maple Butter.

---

4. RICOTTA Pancakes

Ingredients

- 250g full fat ricotta cheese

- 14g coconut flour

- 7g flaxseed meal

- 12g erythritol

- ¾ teaspoon baking powder

- ½ teaspoon salt

- 5 eggs

- ¼ teaspoon xanthan gum

- 12 drops of liquid stevia

- 55g coconut oil

----

16 SERVINGS – Per serving:

89 calories; 7g fat; 4g protein; 2g carbohydrate; 1g dietary fiber; 1g net carbohydrate

Directions

1. First, put a big pan over medium heat; you want it ready when your batter is prepared.

2. Measure everything but the coconut oil into your blender and run until you have a smooth batter.

3. Drip a drop or two of water into your pan; when it skitters around, it's hot enough.

4. Melt 14g of the coconut oil on it, sloshing it around to coat the whole thing.

5. Now, pour the batter out of the blender, into roughly 7.5cm rounds; these are tender and will be easier to turn if you don't make them too big.

6. Fry like any pancakes. Make sure they're quite done on the bottom before turning; look for the top surface to have little holes where bubbles have burst and not filled in and to be starting to look a little dry.

7. Flip and cook the other side. Repeat with the remaining oil and batter.

8. Butter and serve with low-sugar preserves or sugar-free pancakes syrup.

5. Pork Rinds Pancakes

Ingredients

- 55g pork rinds

- 3 eggs

- 60ml pouring cream

- ½ teaspoon baking powder

- 35g erythritol

- ¼ teaspoon liquid stevia

- ½ teaspoon ground cinnamon

- 20g coconut oil, for frying, or more as needed

---

8 PANCAKES – Per serving:

111 calories; 9g fat; 7g protein; 1g carbohydrate; trace dietary fiber; 1g net carbohydrate

Directions

1. Dump your pork rinds in the food processor with the S-blade in place and process until they're reduced to fine crumbs.

2. In a mixing bowl, whisk together the eggs, cream, baking powder, erythritol, liquid stevia, and cinnamon.

3. Now, add the pork rind crumbs, and whisk them in. Let the batter sit for 10 minutes or so. During this time it will thicken up and become gloppy. That's ok.

4. While you're waiting for the gloppification to occur, put your pan over a medium-high heat. You'll want it hot for frying your pancakes.

5. Back to your gloppy batter. Thin it with water if you like-this depends on how thick a pancake you want.

6. Then, fry like any other pancake batter, using the coconut oil as needed. I scoop my batter with an ice cream scoop, so they all come out the same size.

7. Serve with butter and a sprinkle of cinnamon and erythritol or Maple Butter.

6. Peanut Butter Bread

Ingredients

• 95g. almond flour

• 230g vanilla whey protein powder

• 25g erythritol

• 15g. baking powder

• 1 teaspoon salt

• 195g natural peanut butter

• 1 egg

• 1 teaspoon xanthan gum

• ¼ teaspoon of liquid stevia

• 235ml unsweetened coconut milk

———————————

20 SERVINGS – Per serving:

132 calories; 7g fat; 14g protein;

5g carbohydrate; 1g dietary fiber; 4g net carbohydrate

Directions

1. Preheat the oven to 180°C.

2. Generously coat a loaf pan with cooking spray.

3. In your food processor with the S-blade in place, add the almond flour, vanilla whey protein, erytritol, xanthan, baking powder, and salt.

4. Pulse 15 to 20 times, making sure everything is evenly mixed. Add the peanut butter.

5. Pulse 5 to 6 times and then run the processor for a minute or two, scraping down the sides at least once. Turn off the processor.

6. In a cup with a spout, combine the coconut milk, egg, and liquid stevia and use a fork or whisk until well blended.

7. With the food processor running, pour in the coconut milk mixture through the feed tube. When it's all in, scrape down the sides of the processor and process for another 30 seconds or so.

8. Scrape the batter into the prepared loaf pan, smoothing the top. Bake for 1 hour.

9. Cool in the pan for 5 minutes before turning out onto a rack to cool.

10. Serve warm or toasted, slathered with butter.

7. Sunflower Quiche Crust

Ingredients

- 220g raw sunflower seeds

- 50g grated parmesan cheese

- ½ teaspoon salt

- ¼ teaspoon baking powder

- 30ml water

---

12 SERVINGS – Per serving:

118 calories; 10g fat; 5g protein; 4g carbohydrate; 2g dietary fiber; 2g net carbohydrate

Directions

1. Preheat the oven to 180°C. Coat a 25cm pie plate with cooking spray.

2. Put everything but the water in your food processor.

3. Process until it's the consistency of a fine meal. With the food processor still running, drizzle in the water.

4. When you have a soft dough, turn off the machine. Turn the dough out into the pie plate.

5. Use clean hands to press it firmly into an even layer across the bottom and up the sides-you may need to nip a little off here and move it over there to get it even.

6. End the top edge of the plate; don't try to build up a crimped edge like you might with a wheat flour pie crust.

7. Bake for 15 to 17 minutes until very lightly gold. Fill and bake again.

8. Blueberry Muffins

Ingredients

• 95g. almond flour

• 230g. vanilla whey protein powder, divided

• 18g. baking powder

• 1 teaspoon xanthan gum

• ½ teaspoon salt

• 50g erythritol

• 145g. fresh blueberries

• 235ml. unsweetened coconut milk

• 40ml melted butter

• 2 eggs

• 1 teaspoon vanilla extract

• 1/8 teaspoon liquid stevia

• 1 teaspoon lemon zest

---

18 SERVINGS – Per serving:

111 calories; 5g fat; 13g protein; 5g carbohydrate; 1g dietary fiber; 4g net carbohydrate

Directions

1. Preheat the oven to 200°C. Line 18 muffin tins with paper liners.

2. In a mixing bowl, combine the almonds flour, 170g of the vanilla whey, baking powder, xanthan, salt, and erythritol.

3. Stir together well, breaking up any clumped bits of baking powder, until everything is evenly combined.

4. In a separate bowl, toss the blueberries with the remaining of vanilla whey to coat.

5. In a bowl with a pouring lip or a large Pyrex measuring cup, combine the coconut milk, melted butter, eggs, liquid stevia, vanilla, and lemon zest. Whisk together well.

6. Add the liquid ingredients to the dry ingredients all at once and whisk just until everything is dampened and you are sure there are no big pockets of dry stuff left at the bottom of the bowl.

7. Do not try to beat out every lump-over mixing yields an inferior muffin. Quickly stir in the berries.

8. Divide the batter among the muffin cups- an ice cream scoop is useful for keeping them even sizes.

9. Bake for 18 minutes or until a toothpick inserted into the center of a muffin brings out a few moist crumbs. Serve hot, with butter.

9. Pumpkin Muffins

Ingredients

- 32g. almond flour

- 115g. vanilla whey protein powder

- 15g dried egg white powder

- 1 teaspoon baking powder

- ½ teaspoon xanthan gum

- ½ teaspoon salt

- ½ teaspoon ground cinnamon

- ½ teaspoon ground nutmeg

- 125g pumpkin

- 80ml. unsweetened coconut milk

- 30ml melted butter

- 1 egg

- ¼ teaspoon orange extract

- ¼ teaspoon liquid stevia

- 55g chopped pecans

---

12 SERVINGS – Per serving:

118 calories; 7g fat; 10g protein; 4g carbohydrate; 1g dietary fiber; 3g net carbohydrate

Directions

1. Preheat the oven to 200°C. Spray a 12-cup muffin tins with cooking spray.

2. In a mixing bowl, combine all your dry ingredients. Stir them together to evenly distribute ingredients.

3. In another bowl, combine the pumpkin, egg, melted butter, orange extract, coconut milk, and liquid stevia and whisk together

4. Make sure your oven is up to temperature before you take the next step. Pour the wet ingredients into the dry ingredients and with a few swift strokes, combine them.

5. Stir just enough to make sure there are no big pockets of dry stuff; a few lumps are fine.

6. Stir in the pecans quickly and spoon into the prepared muffin tin.

7. Bake for 20 minutes; remove the muffins from the pan to wire rack to cool.

10. Walnut Bread

Ingredients

- 150g walnuts, divided

- 240g shredded coconut

- 2 teaspoons erythritol

- 1 ½ teaspoons baking soda

- ¾ teaspoon salt

- 4 eggs

- 1 teaspoon xanthan gum

- 6 drops of liquid stevia

- 120ml water

- 15ml cider vinegar

———————

20 SERVINGS – Per serving:

140 calories; 12g fat; 5g protein;

5g carbohydrate; 3g dietary fiber; 2g net carbohydrate

Directions

1. Preheat the oven to 180°C. Line a loaf pan with no stick foil.

2. Spread the walnuts on a shallow baking tray, put them in the oven, and set the timer for 6 minutes.

3. Meanwhile, put the coconut, erythritol, baking soda, xanthan, and salt in your food processor, and start it running. Scrape down the sides every few minutes.

4. When the timer beeps, pull the walnuts out of the oven, and add 100g of them to the mixture in the food processor and then run it again. You want to keep running the food processor until the mixture has the texture of a nut butter.

5. When the coconut-walnut mixture reaches a nut butter consistency, add the flaxseed meal and run the processor, scraping down the sides once or twice, until it's well blended.

6. Add the liquid stevia, then the eggs, one by one, blending each in thoroughly before adding another. Don't forget to scrape down the sides when needed.

7. In a glass measuring cup, combine the water and cider vinegar. With the food processor running, pour this through the feed tube in three addition, letting each get worked before adding more. Scrape down the sides if needed.

8. Once the water and vinegar are in, you need to work quickly.

9. Add the remaining 50g of walnuts and pulse a few times to chop them in; you want there to be chunks of walnut in your finished bread.

10. Scrape the dough into your prepared loaf pan, distributing it evenly, and smooth the top.

11. Bake for 75 minutes until it's pulled away from the sides of the pan and sounds hollow when you thump it with a finger. Turn out onto a rack to cool.

11. COCONUT ALMOND FLAX Bread

Ingredients

- 160g shredded coconut

- 50g almond flour

- 115g vanilla whey protein powder

- 55g flaxseed meal

- 9g xanthan gum

- 6 drops liquid stevia

- 1 ½ teaspoon baking soda

- ½ teaspoon salt

- 120ml water

- 30ml cider vinegar

- 4 eggs

---

20 SERVINGS – Per serving:

105 calories; 7g fat; 8g protein; 5g carbohydrate; 3g dietary fiber; 2g net carbohydrate

---

DIRECTIONS

1. Preheat the oven to 180°C. Grease a standard loaf pan (21.6 x 11.4). Now, line it with aluminum foil or baking paper.

2. In your food processor with the S-blade in place, combine the coconut, almond flour, vanilla whey protein, flaxseed meal, xanthan, liquid stevia, baking soda and salt.

3. Run the processor until everything is ground to a fine meal. Scrape down the sides and run the processor some more.

4. While that's happening, in a glass measuring cup, combine the water and the cider vinegar. Have this standing by the food processor.

5. With the food processor running, add the eggs, one at time, through the feed tube.

6. Finally, pour the water and vinegar mixture in through the feed tube. Run just another 30 seconds or so.

7. Pour or scrape the batter into the prepared loaf pan.

8. Bake for 1 hour and 15 minutes. Turn out onto a wire rack to cool.

12. Soft Bread

Ingredients

• 340g cream cheese, at room temperature

• 55g butter

• 60ml MCT oil

• 4 eggs

• 60ml cream

• 380g plain whey protein powder

• 1 teaspoon xanthan gum

• 3 drops liquid stevia

• 12g baking powder

• ½ teaspoon salt

• ½ teaspoon baking soda

• ¼ teaspoon cream of tartar

———————

20 SERVINGS – Per serving:

201 calories; 14g fat; 17g protein; 2g carbohydrate; trace dietary fiber; 2g net carbohydrate

Directions

1. Preheat the oven to 170°C. Coat a 23 x 13 loaf pan with cooking spray or line with a non-stick foil.

2. Put the cream cheese and butter in a microwaveable bowl and nuke for 1 minute on high.

3. Add the MCT oil to the cream cheese and butter and use your electric mixer to beat them until well blended, scraping down the sides of the bowl as needed.

4. Now, beat in the eggs, one at time, incorporating one thoroughly before adding the next. Beat in the liquid stevia and cream.

5. Combine all the dry ingredients in another bowl. Stir them together until everything is evenly distributed.

6. Using a spoon rather than your mixer, stir the dry ingredients into the cream cheese mixture, adding about 85g at time and stirring each addition in before adding more.

7. Pour or scrape the batter into the prepared loaf pan. Bake for 45 minutes or until golden brown.. Turn out onto a wire rack to cool.

8. Store in a plastic bag in the refrigerator. Or, if you're not likely to eat it up quickly, slice it all, wrap it in a plastic bag, and freeze. You can remove and thaw just a slice or two at time.

13. Pizza Crust

Ingredients

• 100g pork rinds

• 230g shredded mozzarella cheese

• 20g shredded parmesan cheese

• 1 clove of garlic, crushed

• 4 eggs

• ½ teaspoon baking powder

• ½ teaspoon salt

• 225g cream cheese, softened

---

16 SERVINGS – Per serving:

151 calories; 12g fat; 10g protein; 1g carbohydrate; trace dietary fiber; 1g net carbohydrate

Directions

You can have pizza. Just top this with pizza sauce, cheese, and your favorite toppings and run it back into the oven for 15 to 17 minutes or until golden and bubbly.

1. If you're going to use your pizza dough right away, preheat your oven to 220°C and line 2 baking sheets with baking paper.

2. Put the pork rinds in your food processor and grind them into fine crumbs. Transfer to a bowl and reserve.

3. Now, put the mozzarella and parmesan in your food processor, along with the garlic, salt, and baking powder, and process until the cheese is finely ground.

4. Add the cream cheese and process for 20 seconds or so. Now, add the eggs, one at time, working each one in well before adding the next. At this point, you will have a soft and sticky mass of dough.

5. If you want to use it right away, divide the ball in two. Place one on a baking paper lined baking sheet.

6. Coat your clean hands with oil and pat or press the dough out quite thin until you have a circle between 25 and 30cm in diameter. Repeat with the second ball of dough.

7. If you make your dough in advance, make your dough balls, put them in a snap-top container or plastic bag, and refrigerate for several hours. I actually left mine in the fridge for 48 hours.

8. Unsurprisingly, the refrigerated dough will be stiffer and less sticky than it was straight out of the food processor. I found I could roll it out with my rolling pin, directly on baking paper. Again, you're going for 25 to 30cm in diameter.

9. Bake your crust for 20 minutes until golden. At this point, top and bake them just as you would any prepared pizza crust.

---

14. SUNFLOWER PARMESAN Crackers

Ingredients

- 220g sunflower seeds

- 80g shredded parmesan cheese

- 1 egg white

- ¼ teaspoon baking powder

- 1 teaspoon xanthan gum

- ½ teaspoon salt, plus more for sprinkling

- 15ml water

---

72 CRACKERS – Per serving:

22 calories; 2g fat; 1g protein;

1g carbohydrate; trace dietary fiber; 1g net carbohydrate

Directions

1. Preheat your oven to 180°C.

2. Put the sunflower seeds, parmesan, baking powder, xanthan, and salt in your food processor and process until the sunflower seeds are a fine meal.

3. With the motor running, add the egg white and water through the feed tube. Keep it running, stopping and scraping down the sides, until you have an evenly mixed soft dough.

4. Line 3 baking sheets with baking paper. Divide the dough in 3 balls.

5. Put one of the balls on a baking-lined baking sheet and put another piece of baking paper over it.

6. Roll it out as thin as you can get it without holes-you may need to nip off a bit of dough from one place and patch it in another to get it as even as possible.

7. Remove the top layer of baking paper and use a thin, straight-bladed knife to score the dough into square or diamonds.

8. Pressing the knife straight down into the dough works better than drawing it along in a slicing motion, I make mine about the size of Wheat Thins, so that's what the analysis is based on.

9. Repeat with the rest of the dough on the other baking sheets. Sprinkle all your crackers lightly with salt. Bake for about 15 minutes or until golden. You may want to turn the baking sheets end to end or even swap shelves to bake them evenly.

10. Let cool, break apart, and store in a snap-top container.

15. Bacon Cheese Crackers

Ingredients

- 340g bacon

- 200g fontina cheese

- 220g sunflower seeds

- ¼ teaspoon baking powder

- ½ teaspoon xanthan gum

- ½ teaspoon pepper

- 1 egg white

- Salt, for sprinkling

---

86 SERVINGS – Per serving:

46 calories; 4g fat; 2g protein;

1g carbohydrate; trace dietary fiber; 1g net carbohydrate

Directions

1. Preheat the oven to 180°C.

2. Start cooking your bacon by your preferred method.

3. Run the fontina through the shredding disk of your food processor. Transfer to a bowl and swap out the disk for the S-blade.

4. Put the sunflower seeds, baking powder. Xanthan, and pepper in the food processor and process until you have a fine meal.

5. If your bacon isn't crisp, just turn off the processor and wait until it is. (You could separate the egg now if you like.)

6. When the bacon is crisp, turn the processor back on. Break the bacon into 7.5 to 10cm lengths and feed them into the processor while it's running.

7. When the bacon is all worked in, the cheese go in, about one-third at time.

8. When the cheese is worked in, add the egg white. Let the processor run until you have a clump of dough.

9. Line a baking sheet with baking paper. Divide the dough into 3 portions. Shape one into a rough ball and put it in the middle of the baking paper.

10. Cover with a second sheet of baking paper. Now, roll the dough out into as thin and even a sheet as you can. Peel off the top layer of baking paper.

11. Using a thin, straight-bladed knife, score the dough into square, triangles, or diamonds. You'll find it far easier to place the whole edge of the blade along the line you want to cut and press down rather than to draw the blade along in a slicing motion. I make my crackers a little larger than Wheat Thins. Sprinkle lightly with salt.

12. Bake for 17 to 18 minutes or until browned. These crackers are better a little overdone than a little underdone. Repeat with each of the remaining balls of dough.

13. Store in a snap-top container for the roughly 36 hours it will take for these to evaporate. They have a way of disappearing.

---

16. PORK RIND Crumbs

Ingredient

• 100g plain pork rinds

---

10 SERVINGS – Per serving:

54 calories; 3g fat; 6g protein;

0g carbohydrate; 0g dietary fiber; 0g net carbohydrate

Directions

1. Dump the pork rinds into your food processor. Process until they are crumbs. Store in a tightly covered container in the refrigerator. That is all.

2. With the processor running, add the oils slowly, in a stream about the diameter of a pencil lead.

3. When it's all worked in, you're done. Scrape it into an old 1l mayonnaise jar and stash it in the refrigerator.

17. Italian Crumbs

Ingredient

- 1 recipe Pork Rind Crumbs (above)

- ¾ teaspoon dried parsley

- ½ teaspoon granulated garlic

- ½ teaspoon onion powder

- ¼ teaspoon oregano

---

10 SERVINGS – Per serving:

55 calories; 3g fat; 6g protein;

trace carbohydrate; trace dietary fiber; 0g net carbohydrate

Directions

If you've been using Italian-seasoned bread crumbs in meatballs or to bread things, try these! They're good in meatloaf, too.

1. Just add the seasonings as you're processing your bag of pork rinds. Store in an airtight container in the fridge, if you're keeping them longer than a week or two.

Chapter 14

SNACKS

1. STUFFED Eggs

Ingredients

- 12 hardboiled eggs

- 45g mayonnaise

- 15g sour cream

- 4g minced parsley

- 4g minced fresh tarragon

- 1 spring onion, including the crisp part of the green, cut into 2.5cm lengths

- 15ml lemon juice

- 1 teaspoon salt

- 2 teaspoons Dijon mustard

- 1 anchovy fillet

- 1 clove of garlic, crushed

- 1 avocado, ripe

- Salt and pepper to taste

---

24 SERVINGS

Per serving:

67 calories; 6g fat; 3g protein;

1g carbohydrate; trace dietary fiber; 1g net carbohydrate

Directions

1. Peel your eggs and halve them, turning the yolks out into a mixing bowl and arranging the whites on a platter.

2. With the S-blade in place in your food processor, add the mayonnaise, sour cream, parsley, tarragon, spring onion, lemon juice, Dijon mustard, anchovy, and garlic.

3. Process until the anchovy vanishes and the herbs are finely chopped.

4. Halve the avocado, remove the pit, and spoon half of the flesh into the food processor. Process again until it's blended in and the mixture is creamy.

5. Turning to those yolks, grab a fork and mash them quite well.

6. Now, add the mixture from the food processor and continue mixing and mashing until it's all creamy and well blended. Spoon in the rest of the avocado in smallish spoonful.

7. Mash and stir until it's blended in, but you still have some nice hunks of avocado.

8. Stuff your egg whites with this fantastic stuff. You'll really have to overstuff them because there will be plenty of yolk mixture. Pile it high.

9. Garnish with more herbs if you like and serve immediately.

---

2. PARMESAN CRISps

Ingredients

• 50g grated parmesan cheese

Do use fresh parmesan, even if it means you have to grate your own.

---

8 SERVINGS – Per serving:

23 calories; 2g fat; 2g protein; trace carbohydrate; 0g dietary fiber; trace net carbohydrate

Directions

1. Preheat the oven to 200°C. Line a baking sheet with baking paper.

2. Put a tablespoon of parmesan on the paper, patting each little heap down a little and spacing them at least 1.5cm apart. Bake for 4 to 5 minutes.

There are also wonderful made in the microwave: coat a microwaveable plate with cooking spray, arrange your little piles of parmesan, and nuke on high for about 1 minute to 75 seconds, depending on the power of your microwave. It's a different texture, but I actually prefer it to the baked; it has the same great flavor.

3. Rosemary Walnuts

Ingredients

- 300g walnuts

- 60ml olive oil

- 3g finely minced fresh rosemary

- 12g erythritol

- 2 teaspoons salt

- 1 teaspoon pepper

---

12 SERVINGS – Per serving:

230 calories; 22g fat; 8g protein; 4g carbohydrate; 2g dietary fiber; 2g net carbohydrate

Directions

1. Preheat the oven to 180°C.

2. Spread the walnuts in a 23 x 33cm baking pan.

3. Combine the olive oil and rosemary in a small bowl. Let it sit for 10 minutes for the flavor to infuse the oil.

4. While that's happening, stir together the erythritol, salt, and pepper.

5. Pour the olive oil and rosemary over the walnuts and stir until they're all evenly coated.

6. Stir in the erythritol mixture in three additions, sprinkling each over the nuts and stirring it in well before adding more. Spread the nuts evenly in the pan.

7. Roast for 20 minutes, stirring two or three times. Let cool and store in an airtight container.

4. Spiced Walnuts

Ingredients

- 200g walnuts

- 28ml MCT or coconut oil

- 35g erythritol

- 1 teaspoon salt

- 1 teaspoon ground cumin

- ½ teaspoon ground coriander

- ¼ teaspoon cayenne

240G – Per serving:

221 calories; 21g fat; 8g protein;

4g carbohydrate; 2g dietary fiber; 2g net carbohydrate

Directions

1. Preheat the oven to 120°C. If you like, line a 23 x 33cm baking pan with foil,.

2. Put the walnuts and the oil in the pan-if you're using coconut oil, melt it first by putting it in the pan and sliding it into the oven for a few minutes. Either way, stir until the walnuts are evenly coated with the oil.

3. Measure everything else into a small dish and stir together. Now, sprinkle this mixture over the walnuts, just a little at a time, stirring well between each addition.

4. When all the spice mixture is in and your nuts are evenly coated, spread them evenly in the pan and put them in to roast. Set the timer for 12 minutes.

5. When the timer beeps, pull your walnuts out, stirring and turning them over, and then spread them out again.

6. Put them in for another 12 minutes. Do this twice more-the stirring and turning and spreading out-for a total of 48 minutes baking time. (This is not crucial, an extra minute or two is not a big deal.)

7. Let the nuts cool in the pan before transferring to an airtight container for storage.

5. Pizza Dip

Ingredients

- 340g cream cheese, at room temperature

- 1 tablespoon Italian seasoning

- 3 cloves of garlic

- 170g pizza sauce

- 115g shredded mozzarella cheese

- 40g shredded parmesan cheese

- 2 tablespoons olive oil

———————

6 SERVINGS – Per serving:

348 calories; 32g fat;

11g protein; 6g carbohydrate;

trace dietary fiber; 6g net carbohydrate

Directions

1. Preheat the oven to 180°C. Coat a 20 x 20cm baking pan with cooking spray.

2. With the S-blade in place in your processor, add the cream cheese, Italian seasoning, and garlic. Process until the garlic is fully pulverized, about 3 minutes.

3. Spread the cream cheese mixture evenly in the prepared pan. Spread the pizza sauce over that, then the mozzarella, and finally the Parmesan. Drizzle the olive oil over the whole thing, as evenly as possible.

4. Bake for 25 to 30 minutes.

Serve hot.

6. Tuna Tapenade

Ingredients

- 170g canned tuna in olive oil, lightly drained

- 25g pitted black olives

- 25g pitted green olives

- 2 anchovy fillets

- 17g capers, drained and rinsed

- 15ml lemon juice

- 1 teaspoon Dijon mustard

- ½ teaspoon dried basil

- 1 pinch of pepper

- 1 clove of garlic, chopped

- 28ml olive oil

- 28g mayonnaise

- 20g diced red onion

---

6 SERVINGS – Per serving:

149 calories; 12g fat; 9g protein; 2g carbohydrate; 1g dietary fiber; 1g net carbohydrate

Directions

Just add everything to your food processor and pulse until the olives are chopped, but not pureed-you want a slightly rough texture.

7. Smoked Salmon Mousse

Ingredients

- 115g cream cheese, softened

- 2 spring onions

- 4g minced fresh dill

- 4g minced fresh parsley

- 1 teaspoon prepared horseradish

- 225g smoked salmon

- 60ml cream

- ¼ teaspoon hot sauce (Tabasco)

- Salt and pepper to taste

---

6 SERVINGS – Per serving:

147 calories; 12g fat; 9g protein;

1g carbohydrate; trace dietary fiber; 1g net carbohydrate

Directions

1. Simply put everything but the salt and pepper in your food processor and process until it's smooth and creamy.

2. Add salt and pepper to taste and process again for just a moment or two to mix it in.

3. Scoop it into a pretty bowl and chill for a few hours to let the flavors blend.

4. Serve with celery and cucumber for dipping.

8. Smoked Salmon Rolls

Ingredients

- 225g whipped cream cheese

- 30g sour cream

- 2 spring onions, finely minced

- 4g minced fresh parsley

- 55g red caviar

- 340g smoked salmon, thinly sliced

---

6 SERVINGS – Per serving:

228 calories; 18g fat;

15g protein; 3g carbohydrate; trace dietary fiber; 3g net carbohydrate

Directions

1. In a mixing bowl, work the cream cheese, sour cream, minced spring onions, and parsley together until well blended.

2. Gently fold in the caviar, trying not to break the eggs.

3. Spread the mixture on slices of salmon and then roll each up and cut into 4cm lengths. Arrange the pretty pinwheels on a lettuce-lined plate.

Chapter 15

MAINS

1. CHICKEN WITH MUSHROOMS Sauce

Ingredients

• 8 slices of bacon

• 2 very large skinless chicken breast (900g)

• 120g Porcini, Portobello, and Button Mushrooms in Cream

• 120ml pouring cream

• 30g cream cheese, cut into several chunks, at room temperature

• 115g sour cream

• Xantan gum

• Paprika

———————

4 SERVINGS – Per serving:

582 calories; 36g fat;

57g protein; 4g carbohydrate;

trace dietary fiber; 4g net carbohydrate

Directions

1. Preheat the oven to 150 °C. Coat a 20 x 20cm baking dish with cooking spray.

2. Lay your bacon on a microwave bacon rack or in a Pyrex pie plate. Microwave for 5 minutes, you're pan cooking it. It should still be limp when it comes out.

3. Cut your chicken breasts in half so you have 4 portions. Wrap each portion in 2 bacon slices, covering as much of the surface as you can. Tuck the ends underneath to hold them in place as you lay them in the prepared pan.

4. In a saucepan over medium-low heat, combine the Porcini, Portobello, and Button Mushrooms in Cream with the pouring cream and cream cheese.

5. As it warms, use a spoon or whisk not only to stir, but also to continue smooching and breaking up the cream cheese to melt it in entirely.

6. When the cream cheese is all melted in, add the sour cream and stir it in. Spoon the sauce over the bacon-wrapped breasts and bake for 1 hour.

7. When the hour is up, use a spatula to remove the chicken to plates. You will notice the sauce looks a mess. Do not panic. When this happened to me, I realized instantly that it was because my homemade sauce lacked the thickening power of the tremendous quantity of cornstarch in canned cream of mushroom soup.

Here's what you do: Grab a whisk and your xanthan gum. Whisking madly, sprinkle in just enough xanthan to bind the sauce again.

8. Spoon over the chicken. Sprinkle a little paprika over it for expression, and you're done.

2. Chicken Stroganoff

Ingredients

- 455g boneless, skinless chicken thighs

- 170g mushrooms

- 1 shallot

- 45g butter

- 1 clove garlic, crushed

- 60ml dry white wine 60ml chicken broth

- 1 teaspoon chicken stock granules

- 115g sour cream

- Xanthan gum (optional)

- Salt and pepper to taste

---

4 SERVINGS – Per serving:

285 calories; 21g fat;

18g protein; 4g carbohydrate; trace dietary fiber; 4g net carbohydrate

---

DIRECTIONS

1. Dice your chicken into 2.5cm cubes. Chop your mushrooms. Mince your shallot.

2. Put your large heavy pan over medium heat and add the butter. When it's melted, swirl it to cover the whole bottom of the pan and then throw in the chicken, mushrooms, shallot, and garlic.

3. Sauté the whole thing, using a spatula to turn everything over now and then, until most of the pink is gone from the chicken and the mushrooms have softened and changed color.

4. Add the wine, chicken broth, and chicken stock granules. Stir it up and turn the heat down to medium-low. Now, cover it with a lid-leave a crack for steam to escape-and let it simmer for 15 minutes.

5. After 15 minutes, uncover and let it simmer for another 5 to 10 minutes until the liquid has cooked down to 3mm in the bottom of the pan.

6. Turn the heat to its lowest setting-once you add the sour cream, you do not want it to boil. Stir in the sour cream, if your sauce seems a trifle thin to you, add a sprinkle of xanthan gum, but it shouldn't need much. Season with salt and pepper to taste.

3. Chicken With Raspberry Sauce

Ingredients

- 4 chicken thighs

- 15ml olive oil

- 1 tablespoon butter

- 1 clove garlic, crushed

- 120ml dry white wine

- 15ml raspberry vinegar

- 18 drops of liquid stevia

- 1 teaspoon chicken stock granules

- ½ teaspoon spicy mustard

- 40g chopped onion

- 30g raspberries

- Salt and pepper

---

4 SERVINGS – Per serving:

285 calories; 21g fat;

17g protein; 3g carbohydrate;

1g dietary fiber; 2g net carbohydrate

Directions

1. Sprinkle the chicken all over with salt and pepper.

In your large heavy pan over medium melt the butter with the olive oil. When the pan is hot, brown the chicken all over. You want it a nice golden shade.

2. While the chicken is browning, stir together the wine, raspberry vinegar, liquid stevia, garlic, chicken stock granules, and mustard, stirring until stock granules is dissolved.

3. When the chicken is a pretty brown, remove it to a plate for a minute. Throw the onion in the pan and sauté for just a minute or two.

4. Pour in the wine mixture and stir it around with the onions. Put the chicken back in, skin-side up. Cover the pan, turn the heat to low, and let simmer for 20 minutes.

5. Add the raspberries, re-cover the pan, and let it cook for another 5 minutes.

6. Plate the chicken and stir the sauce, mashing the berries a little as you do, you want them to flavor the sauce, but you should still be able to see some bits of berry on your plate. Season with salt and pepper to taste and spoon the sauce over the chicken.

4. Basic Roast Duck

Ingredients

- 1 whole duck, about (2.7kg)

- 1medium onion

- 1 clove garlic, cut in half

- 1large celery stalk

- 1large carrot, peeled

- Fresh rosemary

---

6 SERVINGS – Per serving:

1334 calories; 129g fat;

38g protein; 3g carbohydrate;

1g dietary fiber; 2g net carbohydrate

Directions

1. Preheat the oven to 230 °C.

2. Prick your duck's skin all over, especially the areas with a thick layer of fat underneath. Don't pierce the meat, just the skin. Prick liberally-you're letting the excess fat out so your duck won't be greasy.

3. You can use the tip of a sharp knife for this or the tines of a carving fork; I find my dinner forks are not sharp enough for this job.

4. Cut the onion, celery, and carrot into chunks. Stuff these into the body cavity. Put a few sprigs of rosemary too.

5. You can rub your duck with a cut clove of garlic too. Truss your duck-tuck the wing tip underneath and tie the legs together. Put it on a rack in a roasting pan.

6. When the oven's up to 230 °C, put the duck in, turning the temperature down to 180 °C.

7. Now roast your duck for 20 minutes for each 450g, about 2 hours for a 2.7kg. Every 30 minutes or so, pull your duck out and repeat the pricking of the fatty areas.

8. When your duck is done, remove it to a platter and let it rest for 10 to 20 minutes. In the meanwhile, take the rack out of the roasting pan and pour off the duck fat to use for cooking.

9. Pull the veggies out of the body cavity and discard.

5. Pan-Seared Duck Breast With Plum Sauce

Ingredients

• 4 duck breast

• ½ teaspoon salt

• ¼ teaspoon pepper

• ¼ teaspoon dry mustard

• 1 pinch of ground rosemary

• 1 tablespoon bacon grease

• 30ml olive oil

• 1 recipe Plum Sauce

---

4 SERVINGS – Per serving:

643 calories; 62g fat;

15g protein; 6g carbohydrate;

1g dietary fiber; 5g net carbohydrate

Directions

1. Preheat the oven to 190 °C.

2. Using a very sharp, thin-bladed knife, score the skin on each duck breast into 4cm diamonds.

3. Mix together the salt, pepper, dry mustard, and rosemary in a small bowl. Rub the breast with the seasoning mixture.

4. Put your big, heavy pan over medium-high heat. Let it get good and hot and then add the bacon grease and olive oil. Swirl them together and let them heat for a minute.

5. Then add the duck, skin-side down. Let it cook without disturbing it until the skin is brown and crunchy, about 5 minutes.

6. Now, put the pan in the oven and let it cook for another 12 minutes or so. Timing will depend a bit on how thick your breasts are and how rare you like your duck.

7. Plate the duck breasts skin-side up and let rest for 5 minutes. Slice and serve with the Plum Sauce.

6. Fish With Wine, Lemon, And Olives

Ingredients

• 340g cod fillet

• Salt and pepper

• 30g butter

• 30ml olive oil

• ½ of a lemon

• 15g minced parsley

• 30ml dry white wine

• 20 pitted kalamata olives

---

2 SERVINGS – Per serving:

475 calories; 36g fat;

31g protein; 5g carbohydrate;

Trace dietary fiber; 5g net carbohydrate

Directions

1. Divide your fish in portions, with an eye to thickness-you want to put the ticker pieces in to cook first. Season with salt and pepper on both sides.

2. Use a non-stick pan or spray your pan with cooking spray. Put it over medium-low heat and add the butter and olive oil.

3. Swirl them together as the butter melts.

4. When the fat is hot, lay the thick pieces of fish in it-my thickest pieces where about 2 cm. Set the timer for 2 minutes. While that's happening, quarter your lemon and flick out the seeds.

5. When the timer beeps, turn the pieces of fish already in the pan and add the thinner pieces to the pan. Set the timer again for 2 minutes.

6. By the time the second 2 minutes is up, the thick pieces should be flaky clear through; remove from the pan and plate them.

7. The thinner pieces may be done, too, but I like to flip them and give them just a little heat on the other side so they get just a little gold. Plate those, too. Place in a low oven to keep warm or tent with foil.

8. Throw the parsley and wine into the pan and squeeze in the lemon juice. As that's cooking down a little, add the olives; you want to just warm them through.

9. When the sauce has reduced a little-just a minute or two-divide the olives between the plates and then pour the sauce over them. Serve immediately.

7. Artichoke And Clam Risotto

Ingredients

- ½ of a large head of cauliflower

- 235ml water

- 900g small clams, in the shell but scrubbed

- 2 cloves garlic, crushed

- 1 shallot, minced

- 45ml extra virgin olive oil

- 240ml clam juice, no-carb

- ½ of a lemon

- 8g minced parsley

- 235ml dry white wine

- 280g frozen artichoke hearts, thawed and drained

- 6g minced fresh dill

- 1 bay leaf

- Salt and pepper

———————————

4 SERVINGS – Per serving:

358 calories; 13g fat;

33g protein; 17g carbohydrate;

6g dietary fiber; 11g net carbohydrate

Directions

1. Turn the cauliflower into cauli-rice as described on page... While it's cooking prepare the clams.

2. Put the water in a large saucepan and bring to a boil. Add the clams to the boiling water.

3. Cover the pot and let them cook for about 5 minutes or until opened. Remove with a slotted spoon, discarding any clams that have no opened.

4. Strain the water from the clams and reserve. Somewhere in here, your cauli-rice will be done steaming-uncover it or you'll have unappealing mush.

5. Put your saucepan back over medium-low heat. Add the olive oil, then add the garlic and shallot, and sauté them for 3 to 4 minutes.

6. Now, add the strained clam broth, clam juice, wine, and bay leaf. Bring this to a simmer. Reduce it to one-fourth its original volume, about 10 to 15 minutes. While that's happening, remove the clams from their shells, discarding the shells.

7. Quarter your artichoke hearts, too.

8. When your liquid has reduced nicely, remove the bay leaf. Now, stir in the cauli-rice, clams, artichoke hearts, parsley, and dill. Heat everything through and season with salt and pepper to taste.

8. Pan-Fried Salmon With Spinach And Lime-Vodka Cream Sauce

Ingredients

- 340g salmon fillet

- 30g butter

- 1 shallot, minced

- 120g fresh baby spinach

- 160ml fresh cream

- 30ml vodka

- 60ml lime juice

- Salt and pepper

---

2 SERVINGS – Per serving:

630 calories; 47g fat;

38g protein; 8g carbohydrate;

2g dietary fiber; 6g net carbohydrate

Directions

1. If your large heavy pan isn't non-stick, coat it with cooking spray. Put it over medium-heat.

2. Season the meaty side of the salmon fillets with salt and pepper. When the pan is hot, throw in the butter. As soon as it's melted, add the salmon. Give it about 4 minutes-you want it turning glod on the meaty side. Remove the salmon to a plate.

3. Throw the shallot in the pan and saute it for 2 to 3 minutes. Add the spinach and sauté just until it's going limp. Stir in the cream, vodka, and lime juice. Use your spatula to make spaces for your salmon. Lay the fillets in the sauce skin-down.

4. Cover the pan with a tilted lid. Let the whole thing cook for 4 to 5 minutes until your salmon looks cooked through at the thickest spot.

5. Divide the spinach between 2 plates, making 2 beds, and place a fillet, skin-side down, on each.

6. Turn up the heat under the pan and reduce the sauce for another minute and then divide between the servings, pouring it over the salmon.

9. Fish In Coconut-Chilli Sauce

Ingredients

- 910g red snapper fillet
- 30ml lemon juice
- 1 teaspoon garlic powder
- 1 teaspoon salt
- ¼ teaspoon pepper
- 410ml unsweetened coconut milk
- 2 teaspoons grated ginger, divided
- 1 medium green chilli pepper
- Salt and pepper
- 55g coconut oil, divided
- ¼ of a medium red onion, finely diced
- 2 cloves of garlic, crushed
- 2g minced coriander

---

4 SERVINGS – Per serving:

520 calories; 37g fat;

43g protein; 7g carbohydrate;

trace dietary fiber; 7g net carbohydrate

Directions

1. Place the fish in a shallow dish or pan-glass or stainless steel.

2. Rub it all over with the lemon juice. Mix together the garlic powder, salt and pepper and sprinkle evenly over the fish.

3. Stick the dish in the fridge and let your fish marinate for a least 30 minutes, and longer is better. Turn the fish over when you're in there grabbing a sparkling water.

4. Make your sauce next. Put your coconut milk in a saucepan over medium-low heat. As it's warming, add 1 teaspoon of the ginger.

5. Remove the stem, seeds and pith from your chilli pepper. Wash your hands thoroughly with soap and water before you do anything else.

6. Bring the coconut milk to a simmer and then turn it down to just below a simmer and let cook for 15 minutes. Add salt and pepper to taste and turn the heat to its very lowest setting.

7. Go grab your fish out the fridge. Coat your large heavy pan with cooking spray-and put it over medium heat.

8. Melt 30g of the coconut oil and throw in the onion, garlic, and chilli. (Use your spatula to transfer the chilli from your cutting board, or you'll have to go wash your hands again.)

9. Saute for 2 or 3 minutes. Scoop them out with your spatula and stir them into your sauce, leaving the coconut oil in the pan.

10. Add the remaining 25g of coconut oil to the pan. Pat your fish fillets dry with paper towel and throw them in. Let them fry for 3 to 4 minutes per side. Add the sauce and let the fish simmer in it for another 6 to 8 minutes.

11. Plate the fish with the sauce, sprinkling coriander over each serving and dill. Heat everything through and season with salt and pepper to taste.

10. Bacon Scallops

Ingredients

- 4 slices of bacon

- 30g butter

- ¾ teaspoon smoked paprika

- 680g scallops

- 45ml clam juice,

- no-carb

- ¼ teaspoon salt

- 1 pinch of pepper

---

3 SERVINGS – Per serving:

319 calories; 14g fat;

41g protein; 6g carbohydrate;

trace dietary fiber; 6g net carbohydrate

Directions

1. Put your large, heavy pan over medium heat. Use your kitchen shears to snip in the bacon and cook the bacon bits until they're crisp.

2. Scoop them out with a slotted spoon and reserve on a plate, leaving the bacon grease in the pan.

3. Add the butter to the bacon grease. Stir the paprika into the fat and then add the scallops.

4. Saute until cooked through, about 3 to 4 minutes. Add the clam juice and let the whole thing simmer for another minute or two.

5. Add the salt and pepper, stir in the bacon bits, and you're done. Eat before the bacon goes soggy.

11. Rib-Eye Steak With Wine Sauce

Ingredients

- 15ml olive oil

- 680g rib-eye steak

- 2 shallots, chopped

- 120ml dry red wine

- 120ml water and ½ teaspoon beef stock granules

- 15ml balsamic vinegar

- 1 teaspoon Dijon mustard

- 2g dried thyme

- 45g butter

- Salt and pepper

---

4 SERVINGS – Per serving:

428 calories; 28g fat; 35g protein; 2g carbohydrate; trace dietary fiber; 2g net carbohydrate

Directions

1. Cook your steak as describe in Pan-Broiled Steak (below).

2. In the meanwhile, in a bowl, combine the shallots, wine, beef stock, balsamic vinegar, mustard, and thyme in a measuring cup with a pouring lip. Whisk'em up.

3. When the timer goes off, flip the steak and set the timer again.

4. When your steak is done, put it on a platter. Pour the wine mixture into the pan and stir it around, scraping up the nice browned bits, and let it boil hard.

5. Continue boiling your sauce until it's reduced by a least half. Melt the butter, season with salt and pepper, and serve with your steak.

12. Pan-Broiled Steak

Ingredients

• 680g raw steak, 2.5cm thick

• 15g bacon grease or 15ml olive oil

---

4 SERVINGS – Per serving:

403 calories; 33g fat;

24g protein; 0g carbohydrate;

0g dietary fiber; 0g net carbohydrate

Directions

1. Pour your big, heavy pan (cast iron is best) over highest heat and let it get good and hot. In the meanwhile, you can season your steak if you like.

2. You could top it when done with butter and blue cheese, or sautéed onions and mushrooms, or go for a classic simplicity and just salt and pepper it.

3. When the pan's hot, add the bacon grease and slosh it around and then throw in your steak.

4. Set a timer for 5 to 6 minutes-your timing will depend on your taste and how hot your burner gets. On my stove, 5 minutes per side with a 2.5cm thick steak comes out medium-rare.

5. When the timer goes off, flip the steak and set the timer again. Let the steak rest for 5 minutes before devouring.

---

13. BRIE AND WALNUT Burgers

Ingredients

- 1 medium onion, thinly sliced

- 20g butter, divided

- 30g chopped walnuts

- 8 slices of bacon 120g brie

- 680g ground beef

---

4 SERVINGS – Per serving:

708 calories; 58g fat;

42g protein; 4g carbohydrate;

1g dietary fiber; 3g net carbohydrate

Directions

1. Put a medium size pan over medium-low heat and start sautéing the onions in 15g of the butter.

2. You want to caramelize them-let them get limp, brown, and sweet without burning them. This will take 15 minutes or so.

3. In the meanwhile, put a small pan over medium-low heat and melt the remaining butter.

4. Sauté the walnuts in it just until they smell toasty, about 5 to 7 minutes. Remove from the heat.

5. Cook your bacon however you wish to cook your bacon. Whatever, just get it nice and crisp.

6. Slice your brie into 4 portion, removing the white rind. Slice it thinly enough that you can cover each burger.

7. Okay. Your topping are prepared. Form your ground beef into 4 patties about 1.3cm thick. Put your large, heavy pan over medium heat and let it get good and hot.

8. Throw in your burgers and cook them to your liking-I give mine about 5 minutes per side.

9. When you've flipped your burgers and the second side has just another minute or two to cook, spread 8g of the walnuts on each, and then cover with brie. Let it melt as the burgers finishing cooking.

10. Plate your burgers, top with the bacon and caramelized onions

14. Beef And Mushrooms In Ginger Coconut Milk

Ingredients

- 30g coconut oil

- 340g ground beef

- 220g mushrooms, sliced

- 1 shallot, minced

- 2 cloves of garlic, crushed

- 15g grated ginger

- 160ml unsweetened coconut milk

- 230ml beef broth

- 15ml soy sauce

- Xanthan gum

---

4 SERVINGS – Per serving:

384 calories; 32g fat;

20g protein; 6g carbohydrate;

1g dietary fiber; 5g net carbohydrate

DIRECTIONS

1. Put your large, heavy pan over medium heat and melt the coconut oil. Add the ground beef and start to brown and crumble it.

2. As fat starts to cook out of the beef, add the mushrooms, shallot, garlic, and ginger. Sauté it all together until the beef is browned and the mushrooms softened.

3. Stir in the coconut milk, beef broth, and soy sauce. Turn the heat down so the mixture is just simmering. Let it cook for 5 minutes.

4. Thicken just a little with your xanthan gum and serve.

15. Italian Meatloaf

Ingredients

- 2 medium zucchini

- 1 medium onion

- 45ml olive oil

- 680g ground beef

- 2 cloves of garlic

- 75g grated parmesan cheese

- 1 egg, beaten

- 8g chopped fresh parsley

- 1 teaspoon salt

- ½ teaspoon pepper

---

6 SERVINGS – Per serving:

434 calories; 34g fat;

26g protein; 4g carbohydrate;

1g dietary fiber; 3g net carbohydrate

Directions

1. Preheat the oven to 180 °C.

2. Dice your zucchini and onion, or, if you prefer, put them in your food processor and pulse until chopped medium –fine.

3. Place your large, heavy pan over medium heat and add the olive oil. Sauté the zucchini and onion.

4. When they're half-done, crush in the garlic.

5. Keep sautéing until the onion is translucent, about 7 to 8 minutes total. Remove from the heat and let the vegetables cool for a few minutes.

6. Meanwhile, in a large mixing bowl, combine the ground beef, parmesan, egg, parsley, salt, and pepper.

7. When the vegetable are cool enough to handle safely, throw them in, too. Use clean hands to smooch everything together quite well.

8. Pack the meat mixture into a big loaf pan, smoothing the top with a wet hand.

9. Bake for 75 to 90 minutes, or until the juices run clear or it reach an internal temperature of 71 °C. Let cool for 10 minutes in the pan before serving.

16. Carnitas

Ingredients

- 680g boneless pork shoulder

- 1 teaspoon salt

---

4 SERVINGS – Per serving:

301 calories; 23g fat;

22g protein; 0g carbohydrate;

0g dietary fiber; 0g net carbohydrate

Directions

1. Cut the pork into chunks roughly 4 to 5cm.

2. Put the pork in your big heavy pan-you want a single layer, but the pork cubes can be very close together, that's fine. Cover with water, add the salt, and put over medium-high heat.

3. Bring the water to a boil and then turn the heat down to low-you want to keep the water barely simmering.

4. Now, let your pork cubes simmer. And simmer. And simmer. Mine took a good 3 or 4 hours. That's ok; the simmering make them tender. If you happen to be wandering through the kitchen, turn the cubes over once or twice.

5. Let your pork simmer until the water has completely cooked away. Continue cooking, letting the pork cubes brown in the fat that has collected in the bottom of the pan. When they're crisply brown, they're done.

17. Coffee-Rubbed Pork Chops

Ingredients

- 20g espresso ground coffee beans

- 12g erythritol

- 2 teaspoons salt

- 1 teaspoon chilli powder

- 1 teaspoon hot smoked paprika

- 680g pork chops

- (4 chops)

- 15g bacon grease

---

4 SERVINGS – Per serving:

296 calories; 20g fat;

26g protein; 1g carbohydrate;

trace dietary fiber; 1g net carbohydrate

Directions

1. Mix together the coffee, erythritol, salt, chilli powder, and paprika in a small bowl. Rub both sides of your pork chops liberally with this mixture.

2. Put your biggest heavy pan or possibly two, depending on the size of your pan and the size of your chops, over medium heat. Let them get good and hot.

3. Throw in the bacon grease and slosh it about as it melts to coat the bottom of the pan. Now, throw in your chops.

4. Cooking time will depend on the thickness of your chops. Mine where about 1.3cm and took about 7 minutes per side.

5. You want them done through 71 °C, but not dry out. If your chops are thicker than mine, turn down the heat a bit, and cover the pan you're a tilted lid to reflect heat back at the chops.

6. This way you can cook them through before they're scorched on the outside-browned, good; scorched, bad. I think coleslaw would be perfect side with this, but it's up to you.

18. Pulled Pork

Ingredients

- 1 medium onion, thinly sliced

- 1.4kg pork shoulder

- Salt and pepper

- 120ml chicken broth

- 24g grated ginger

- ¼ teaspoon liquid stevia

- 30ml DaVinci sugar-free pineapple syrup

- 30ml apple cider vinegar

- 15ml soy sauce

- 1 clove of garlic, crushed

- 1 head of cabbage, coarsely chopped

- 1 small onion, diced

———————————

6 SERVINGS – Per serving:

532 calories; 36g fat;

45g protein; 5g carbohydrate;

1g dietary fiber; 4g net carbohydrate

Directions

1. Cover the bottom of your slow cooker with the sliced onion.

2. Use a carving fork to stab your pork all over-really go full-tilt serial killer on the thing. Season it with salt and pepper all over and plunk it on top of the onion.

3. Stir together the broth, ginger, liquid stevia, pineapple syrup, apple cider vinegar, soy sauce, and garlic in a bowl and pour it over the roast.

4. Slap on the lid, set it to low, and forget about it for good 8 to 10 hours.

5. When cooking time is up, use tongs to fish out your pork and place it on a platter.

6. Ladle out about 120ml of the liquid from the pot and pour it over the pork.

7. Stir the cabbage and onion into the liquid left in the pot. Put the lid back on, crank the pot to high, and let cook for 45 minutes.

8. In the meanwhile, use 2 forks to shred the pork, discarding the bones. Cover the pork to keep it moist and keep it in a warm place, such as your oven on its lowest setting.

9. Serve the pork with the cabbage and onion.

10. Throw in the bacon grease and slosh it about as it melts to coat the bottom of the pan. Now, throw in your chops.

Cooking time will depend on the thickness of your chops. Mine where about 1.3cm and took about 7 minutes per side. You want them done through 71 °C, but not dry out.

If your chops are thicker than mine, turn down the heat a bit, and cover the pan you're a tilted lid to reflect heat back at the chops. This way you can cook them through before they're scorched on the outside-browned, good; scorched, bad. I think coleslaw would be perfect side with this, but it's up to you.

19. Grilled Pork Tenderloin

Ingredients

• 120ml MCT oil

• 80ml soy sauce

• 60ml red wine vinegar

• 45ml lemon juice

• 28ml Worcestershire sauce

• 1 clove of garlic, crushed

• 4g chopped parsley

• 9g dry mustard

• 1 ½ teaspoons pepper

• 18 drops of liquid stevia

• 680g pork tenderloin

——————

6 SERVINGS – Per serving:

318 calories; 22g fat;

25g protein; 4g carbohydrate;

trace dietary fiber; 1g net carbohydrate

——————

DIRECTIONS

1. Combine everything but the pork in a medium-size bowl.

2. Put the tenderloin in a 3.8L resalable plastic bag.

3. Pour in the marinade and seal the bag, carefully pressing out the air as you go.

4. Turn the bag a few times to coat and then throw it in the fridge. Let the pork marinate for a least 4 to 5 hours, and all day is fine.

5. When cooking time comes, fire up the grill. Pour the marinade off into a small saucepan, place on the stove, and bring to a boil. Remove from the heat and set aside.

6. Grill your tenderloins over well-ashed coal or a gas grill set on medium for about 15 to 16 minutes or until a meat thermometer stuck in the thickest part registers 71 °C.

7. Baste frequently with the reserved marinade.

8. Serve the remaining marinade as sauce with the pork.

20. Roast Leg Of Lamb

Ingredients

• 2.3kg leg of lamb

• 6 cloves of garlic, thinly sliced lengthwise

• 80ml olive oil

• Juice of ½ a lemon

• 1 ½ tablespoons minced fresh rosemary

• 1 ½ tablespoons minced fresh oregano

• ¼ teaspoon pepper

—————————————

6 SERVINGS – Per serving:

800 calories; 63g fat;

54g protein; 2g carbohydrate;

trace dietary fiber; 2g net carbohydrate

Directions

1. Start several hours before you want to serve your lamb, so you can give it time to marinate.

2. Now, put your lamb on the cutting board, and using a paring knife, go all serial killer on it.

3. Stab it about 5cm apart, all over, on both sides.

4. Now, insert a sliver of garlic into each hole-push them in well. Put your garlic-studded lamb in a Pyrex pan.

5. Mix together the olive oil, lemon juice, minced herbs, and pepper.

6. Rub this mixture evenly over both sides of your lamb. Now, let it sit in the fridge for a least 2 to 3 hours. About 3 hours before dinnertime, preheat your oven to 200 °C.

7. Put the lamb on a rack in a roasting pan. Insert a meat thermometer into the thickest part, but not touching the bone.

8. When the oven is up to temperature, put the lamb in and immediately turn the oven down to 170 °C. Now, let it roast for about 30 minutes per each 450g of weight, a bit less if you prefer your lamb rare. (I like mine medium to medium-well.) Check the meat thermometer to be sure-63 C °to 66 °C will be rare; 71 °C will be well done.

9. Remove to a platter and let the lamb rest for 15 minutes while you make your gravy. Then, carve and serve.

21. Harissa Lamb

Ingredients

- 1.4kg lamb shank (2 shanks)

- 30g harissa

- 2 teaspoons erythritol

- ½ of an onion, diced

- 6 cloves of garlic, crushed

- 235ml chicken broth

- 120ml dry white wine

- 60ml olive oil

- 1 tablespoon smoked paprika

- 15g tomato paste

- 1 teaspoon beef stock granules

- ½ teaspoon ground cinnamon

- 1 bay leaf

- 15ml lemon juice

- ½ teaspoon pepper

- Xanthan gum

———————————

3 SERVINGS – Per serving:

956 calories; 68g fat;

70g protein; 6g carbohydrate;

1g dietary fiber; 5g net carbohydrate

Directions

1. Place your lamb shanks on a plate. Mix together the harissa, erythritol, and pepper in a bowl. Set aside 10g of the mixture and rub the rest all over your shanks.

2. In your large, heavy pan over medium heat, sear the shanks in the olive oil, getting them nice and brown all over. Transfer them to your slow cooker.

3. Add the onion and garlic to the pan and sauté for just a minute or two. Add the chicken broth, wine, paprika, tomato paste, beef stock granules, cinnamon, and reserved harissa mixture.

4. Bring to a boil, stirring to dissolve all the nice browned stuff stuck to the pan. Let it boil hard for 4 to 5 minutes and then pour over the shanks.

5. Throw in the bay leaf, making sure it lands in the liquid. Cover the pot, set it to low, and let it cook for 8 to 10 hours.

6. It's dinnertime. Fish out the lamb shanks with tongs and put them on a platter. Remove the bay leaf and discard.

7. Stir in the lemon juice. Use your xanthan to thicken the sauce to the texture of heavy cream.

8. Carve the shanks into portions and serve with the sauce .

22. LEFTOVER CHICKEN Salad

Ingredients

- 280g diced chicken

- 60g sliced celery

- 1 bunch spring onions, sliced, including the crisp part of the green shoot

- 180g mayonnaise

- 45ml lime juice

- 2 teaspoons dark rum

- ½ teaspoon pepper

- ½ teaspoon salt

- ½ teaspoon ground cinnamon

- ½ teaspoon ground ginger

- ½ teaspoon dried thyme

- ½ teaspoon cayenne

- ½ teaspoon ground cloves

- 18 drops of liquid stevia

- 240g shredded romaine lettuce

- 70g chopped oil-roasted macadamia nuts

---

4 SERVINGS – Per serving:

688 calories; 65g fat; 24g protein; 7g carbohydrate; 3g dietary fiber; 4g net carbohydrate

Directions

1. Put the chicken, celery, and spring onions in a mixing bowl.

2. In a bowl, combine the mayonnaise, lime juice, rum, pepper, salt, cinnamon, ginger, thyme, cayenne, cloves, and liquid stevia, stirring it all up.

3. Pour this dressing over the chicken mixture and stir to coat.

4. Create 4 beds of shredded lettuce and place scoops of the chicken mixture on them.

5. Scatter 2 tablespoons of chopped macadamias over each serving.

23. Buffalo Chicken Salad

Ingredients

• 1 recipe Buffalo Slaw

• 210g diced chicken

---

4 SERVINGS – Per serving:

515 calories; 45g fat;

24g protein; 8g carbohydrate;

3g dietary fiber; 5g net carbohydrate

Direction

Combine the Buffalo Slaw and the chicken in a mixing bowl. That's it. It's a great dish for a hot summer night.

24. Prawns And Avocado Salad

Ingredients

- 900g cooked small prawms, peeled and deveined

- 1 avocado

- 10 spring onions, thinly sliced, including the crisp part of the green shoot

- 160ml vinaigrette

- 1 head of romaine lettuce

————————————

6 SERVINGS – Per serving:

352 calories; 21g fat;

35g protein; 8g carbohydrate;

4g dietary fiber; 4g net carbohydrate

Directions

1. Put the prawns in a big mixing bowl. Peel and pit your avocado and dice it, somewhere between 6 to 13mm big. Put that in the bowl along with the spring onions.

2. Pour on the vinaigrette, and gently stir the whole thing up to coat all the ingredients. Let that sit for a few minutes while you break or cut up the lettuce. Arrange it in beds on 6 serving plates.

3. Now, stir the prawns salad one last time to get up any dressing that's settled to the bottom of the bowl and spoon it out onto the beds of lettuce. Serve immediately.

25. Scallops With Asparagus And Greens

Ingredients

- 45ml rice vinegar, divided

- 30ml lemon juice, divided

- ½ teaspoon dark sesame oil

- 60ml MCT oil, divided

- 10 asparagus spears

- 110g mixed greens

- 6 scallops

- Salt and pepper

---

2 SERVINGS – Per serving:

317 calories; 29g fat;

8g protein; 10g carbohydrate;

4g dietary fiber; 6g net carbohydrate

DIRECTIONS

1. Whisk together 30ml of the rice vinegar with 15ml of the lemon juice, the sesame oil, and 30ml of the MCT oil. Set aside.

2. Snap the end of the asparagus where they want to break naturally.

3. Put them in a microwavable casserole dish with a lid, add a couple of tablespoons (30ml) of water, cover, and microwave on high for just 3 to 4 minutes. Uncover immediately.

4. Divide the asparagus spears between 2 salad plates, fanning them out artistically.

5. Top with the salad greens.

6. Sprinkle the scallops with a little salt and pepper.

7. Spray your big pan with cooking spray and put it over a medium-high heat.

8. Add the remaining 30ml of MCT oil and let it get good and hot. Add the scallops and cook for about 2 to 3 minutes per side or until opaque in the center. Remove to a bowl.

9. Put the remaining 15ml of rice vinegar and remaining 15ml of lemon juice in the pan and boil down until reduced to just a couple of teaspoons. Drizzle over the scallops.

10. Now, drizzle the dressing you made first thing over the salad greens, top each serving with 3 scallops, and serve.

26. Cajun Steak Caesar Salad

Ingredients

- 12g Cajun seasoning

- 340g sirloin steak

- 1 ½ head of romaine lettuce

- ¼ of a red onion, sliced paperthin

- 100g grated pecorino romano cheese

- 175ml Caesar Dressing (page...)

- 1 teaspoon Tabasco sauce

---

3 SERVINGS – Per serving:

684 calories; 54g fat;

35g protein; 11g carbohydrate;

3g dietary fiber; 8g net carbohydrate

Directions

1. Preheat the grill.

2. Sprinkle the Cajun seasoning on both sides of the steak and put the steak in the grill.

3. While the steak is grilling, break up the lettuce into your salad bowl. Add the onion and toss to combine.

4. Coat a microwavable plate with cooking spray and spread the pecorino romano cheese on it. Put it in the microwave for 1 minute on high.

5. Stir the Caesar Dressing and the Tabasco together. Pour over the romaine and onion and toss.

6. Okay, your steak should be done now. Pull it out and slice it thin across the grain. Pile the lettuce on 3 serving plates and top with the sliced steak.

7. Pull the cheese out of the microwave-it will be a crisp sheet. Crumble the crisp cheese over the salads and serve.

Chapter 16

SOUPS

1. CREME OF Courgette

Ingredients

- 1 small onion, diced

- 1 small carrot, shredded

- 4 cloves of garlic, crushed

- 60ml olive oil

- 5 medium zucchini, sliced

- 2g fresh thyme

- 1 teaspoon grated ginger

- 1 teaspoon Greek seasoning

- ¼ teaspoon red chilli flakes

- 1.4l chicken broth

- 350ml cream

- 15ml lemon juice

- Salt and pepper

---

8 SERVINGS – Per serving:

272 calories; 24g fat;

6g protein; 8g carbohydrate;

2g dietary fiber; 6g net carbohydrate

Directions

1. Put your stockpot over medium heat and sauté the onion, carrot, and garlic in the olive oil for 5 minutes. Add the zucchini and sauté for another 4 to 5 minutes.

2. Add the thyme, ginger, Greek seasoning, red chilli flakes, and chicken broth. Bring to simmer and let cook until the vegetable are tender, about 20 minutes.

3. Use your stick blender to puree the vegetable. Blend in the cream and lemon juice. Season with salt and pepper to taste.

2. Salmon Bisque

Ingredients

- 40g butter
- 60g finely diced celery
- 80g minced shallot
- 1.9l fish stock
- 230ml dry white wine
- 1 lemon
- 6g minced dill
- 680g salmon fillet, skinned
- 1 teaspoon paprika
- 350ml cream
- Xanthan gum
- Salt and pepper

---

8 SERVINGS – Per serving:

423 calories; 30g fat;

19g protein; 7g carbohydrate;

1g dietary fiber; 6g net carbohydrate

Directions

1. In your stockpot over low heat and melt the butter and sauté the celery and shallot, without browning, for 5 minutes.

2. Add the fish stock and wine and turn the heat up to medium. Slice the lemon in half, remove any seeds, and the drop the halves into the pot. Stir in the dill.

3. When the stock is warmed, gently lower the salmon into the pot. When the stock reaches a simmer, turn the heat down to hold in there and cook for 20 to 25 minutes.

4. Now, you have a decision to make: do you want chunks of salmon in your finished soup, or do you want to puree the whole thing? If you want chunks, carefully fish out some of your salmon with a slotted spoon and lay it on a plate. Don't worry if some pieces fall off and remain in the soup. You do want to puree a least half of the salmon.

5. Add the cream and use your stick blender to puree everything. You could transfer the soup to your blender instead, but you'd have to work in batches.

6. Obviously, if you don't want chunks of salmon in your finished soup, just puree the whole thing-you don't even need to remove the salmon from the pot if you don't care about scraping off any connective tissue.

7. Use your xanthan gum to thicken the soup a little-not quite as thick a heavy cream. Season with salt and pepper to taste.

3. Black Olive Soup

Ingredients

- 950ml chicken stock, divided

- ½ teaspoon xanthan gum

- 100g minced black olives

- 230ml cream

- 60ml dry sherry

- Salt and pepper

---

6 SERVINGS – Per serving:

200 calories; 18g fat; 4g protein; 3g carbohydrate; 1g dietary fiber; 2g net carbohydrate

Directions

1. Put 120ml of the chicken stock in the blender with the xanthan gum and blend for a few seconds. Pour into a saucepan and add the remaining 830ml of stock and the olives.

2. Heat to simmering and then whisk in the cream. Bring back to a simmer, stir in the sherry, and season with salt and pepper to taste.

4. Cream of Cauliflower

Ingredients

- 120g diced onion

- 90g diced celery

- 40g butter

- 945ml chicken broth

- 300g of frozen cauliflower

- 2 teaspoons of xanthan gum (optional)

- 475ml cream

- Salt and pepper

---

4 SERVINGS – Per serving:

250 calories; 21g fat;

7g protein; 9g carbohydrate;

3g dietary fiber; 6g net carbohydrate

Directions

1. In a large saucepan over low heat, sauté the onion and celery in the butter until they're limp.

2. Add the chicken broth and cauliflower and simmer until the cauliflower is tender, about 20 minutes.

3. Using a slotted spoon, transfer the vegetables to a blender and then pour in broth to cover. Add the xanthan gum if you're using it.

4. Remove the little cap to let steam escape and cover the hole with a kitchen towel. Be careful-hot soup will expand in the blender and splatter.

5. You may need to do this in batches. Puree everything in the blender until smooth.

6. Pour back into the saucepan. Stir in the cream and season with salt and pepper to taste.

5. Spicy Chicken And Mushroom Soup

Ingredients

- 1 leek

- 55g butter

- 225g sliced mushrooms

- 1clove of garlic, crushed

- 2 teaspoons garam masala

- 1 teaspoon pepper

- ¼ teaspoon cayenne

- ¼ teaspoon ground nutmeg

- 950ml chicken broth

- 340g boneless, skinless chicken breast, cut into thin strips

- 235ml cream

- 3g chopped fresh coriander

---

6 SERVINGS – Per serving:

318 calories; 25g fat;

18g protein; 6g carbohydrate;

1g dietary fiber; 5g net carbohydrate

Directions

1. Thinly slice the white, crisp part of your leek, discarding the green top. Wash well to remove all the grit.

2. Melt the butter in your big, heavy pan over a medium heat and sauté the leek with the mushrooms until they both soften.

3. Stir in the garlic, garam masala, pepper, cayenne, and nutmeg and sauté for another minute or two. Transfer to your slow cooker.

4. Pour in the chicken broth and add the cut-up chicken breast. Cover the pot, set to low, and let cook for 6 to 7 hours.

5. When time's up, use a slotted spoon to scoop roughly two-thirds of the solids into your blender or food processor. Add 235ml or so of the broth and puree until smooth.

6. Stir back into the rest of the soup (you may want to rinse out the blender or food processor with a little broth, to get all of the puree).

7. Alternatively, you could use your stick blender right in the pot, but take care to leave some chunks of meat and veggies for texture.

8. Stir in the cream. Re-cover the pot and let the whole thing cook for another 30 minutes. Serve with chopped fresh coriander on top. Or not, if you prefer-it's nice without it, too.

6. Broccoli-Blue Cheese Soup

Ingredients

- 55g butter

- ½ of a medium onion, diced

- 450g frozen broccoli, thawed

- 700ml chicken broth

- 350ml cream

- 110g crumbled blue cheese, divided

- Xanthan gum (optional)

- Salt and pepper

---

4 SERVINGS – Per serving:

573 calories; 54g fat; 15g protein; 10g carbohydrate; 4g dietary fiber; 6g net

Carbohydrate

Directions

1. In a large saucepan over medium heat, melt the butter. Add the onion and sauté until just soft.

2. Add the broccoli and broth. Bring to a simmer, cover, and let it cook for 15 to 20 minutes until the broccoli is soft but not gray. Use your stick blender to blend the broccoli until it's smooth.

3. Stir in the cream and half of the blue cheese, mixing until the cheese is just melted. Thicken a little with your xanthan if you feel it's needed. Season with salt and pepper to taste.

4. Stir in the rest of the blue cheese and serve before it melts.

7. Courgette-Gorgonzola Soup

Ingredients

- 45ml olive oil

- 30g butter

- ½ of a medium onion, chopped

- 900g small zucchini

- 1 teaspoon dried oregano

- 945ml chicken broth

- 110g crumbled Gorgonzola cheese, plus more for garnish (optional)

- Xanthan gum (optional)

- 235ml cream

- Salt and pepper

---

6 SERVINGS – Per serving:

348 calories; 32g fat;

10g protein; 7g carbohydrate;

2g dietary fiber; 5g net carbohydrate

DIRECTIONS

1. In a large pot over medium heat, warm the olive oil and butter together. When the butter is melted, add the onion and sauté for 5 minutes until just turning translucent.

2. While that's happening, slice your zucchini. When the onion's soft, add the zucchini to the pot, along with the oregano.

3. Sauté for 10 minutes, stirring often. Add the chicken broth and bring to a simmer. Let it cook for 20 to 30 minutes until the zucchini is quite soft.

4. Add 145ml of the cream and the Gorgonzola. Use your stick blender to blend the soup smooth and bring back to a simmer.

5. Season with salt and pepper to taste and thicken with your xanthan if you feel it needs it.

6. Serve 15ml of the remaining cream swirled into each dish. Garnish with additional Gorgonzola, if you like.

Chapter 17

HOTSIDE DISHES

1. FAUXtatoes

Ingredients

- ½ of a large head of cauliflower

- 30g cream cheese

- 40g butter

- Salt and pepper

---

3 SERVINGS – Per serving:

159 calories; 15g fat;

3g protein; 5g carbohydrate;

2g dietary fiber; 3g net carbohydrate

Directions

1. Trim the leaves and the very bottom of the steam from your cauliflower and whack it into chunks. Steam until tender; I put mine in a microwave.

2. If you don't have a microwave steamer, a microwavable casserole dish with a lid will work fine; just add a few tablespoons (45ml) of water and cover. Or, you can steam it on the stove.

3. When your cauliflower is tender, drain it very well. This is essential. Now puree your cauliflower; I put mine in a deep, narrow bowl and use my stick blender, but a food processor will do the trick, too.

4. When the cauliflower is getting to the pureed stage, add the cream cheese and butter and work them in. Add salt and pepper to taste. These are fine as is, or with gravy, or just a little steak juice spooned over them.

2. Fauxtato Pancakes

Ingredients

- 1 recipe Fauxtatoes

- 3 eggs

- 9g coconut flour

- 50g lard, bacon grease, (60ml) MCT oil, or olive oil, plus more as needed

––––––––––––

6 SERVINGS – Per serving:

203 calories; 19g fat; 5g protein; 5g carbohydrate; 3g dietary fiber; 2g net carbohydrate

Directions

1. This is darned simple: Put your fauxtatoes in a mixing bowl, add the eggs and coconut flour, and whisk it up.

2. Put your big, heavy pan over medium heat and add the fat-I use lard.

3. When the pan is hot, add the batter by the 60ml. Fry until nicely browned on the bottom, flip, and brown the other side. Serve hot.

3. Cauli-Rice

Ingredients

- ½ of a large head of cauliflower

---

4 SERVINGS – Per serving:

18 calories; trace fat;

1g protein; 4g carbohydrate;

2g dietary fiber; 2g net carbohydrate

Directions

1. Trim the leaves and the very bottom of the stem from your cauliflower and cut the rest into pieces that will fit in the feed tube of your food processor. Run it through the shredding blade.

2. Steam the resulting shreds lightly; I cook mine for about 7 or 8 minutes on high in the microwave. Uncover immediately to avoid mushiness and drain well. Season and use in a million ways.

4. Cauliflower Couscous

Ingredients

- 680g cauliflower, riced

- 85g butter

- 80g pistachio nuts, coarsely chopped

- 12g Greek seasoning 30ml lemon juice 10g chopped garlic

- ¼ teaspoon pepper

---

8 SERVINGS – Per serving:

163 calories; 14g fat; 4g protein; 8g carbohydrate; 3g dietary fiber; 5g net carbohydrate

Directions

1. Steam the cauli-rice briefly.

2. Uncover the cauliflower immediately and drain if needed.

3. In a wok or big darned pan, over medium heat, melt the butter. Add the cauli-rice, then everything else. sauté, stirring often, for 4 to 5 minutes.

---

5. VENETIAN Rice

Ingredients

- ½ of a head of cauliflower

- 15ml olive oil

- 15g butter

- 70g sliced mushrooms

- 3 anchovy fillets, minced

- 1 clove of garlic, crushed

- 15g grated parmesan cheese

---

3 SERVINGS – Per serving:

136 calories; 11g fat;

6g protein; 7g carbohydrate;

3g dietary fiber; 4g net carbohydrate

Directions

1. Run the cauliflower through the shredding blade of your food processor.

2. Put it in a microwavable casserole dish with a lid, add a couple of tablespoons (28ml) of water, cover, and nuke on high for 5 to 6 minutes. When it's done, uncover immediately.

3. Combine the olive oil and butter in your big heavy pan over medium heat, swirling together as the butter melts.

4. Add the mushrooms and sauté until they're soft and changing color. If your mushroom slices are quite large, you may want to break them up a bit with the edge of your spatula as you stir.

5. When the mushrooms are soft, stir in the minced anchovies and garlic.

6. Add the cauli-rice, untrained-that little bit of water is going to help the flavors blend. Stir well to distribute all the flavors.

7. Stir in the parmesan and serve.

6. CREAMED SPINACH With Browned Butter

Ingredients

• 55g butter

• 560g frozen chopped spinach, thawed and well drained

• 120ml pouring cream

• 25g grated parmesan cheese

• ¼ teaspoon ground nutmeg

• Salt and pepper

---

8 SERVINGS – Per serving:

131 calories; 12g fat; 3g protein; 3g carbohydrate; 2g dietary fiber; 1g net carbohydrate

Directions

1. In a large saucepan over medium heat, melt the butter.

2. Let it cook until it turns golden brown and nutty smelling, about 6 to 8 minutes. Don't let it burn.

3. Mix in the spinach and cream and bring to a simmer. Let cook for 5 minutes.

4. Stir in the parmesan and nutmeg and season with salt and pepper to taste.

7. Zoodles

Ingredients

- 3 small zucchini

- Salt

- 45ml olive oil

- 1 clove of garlic, crushed

- 2 teaspoons oregano

---

4 SERVINGS – Per serving:

113 calories; 10g fat;

2g protein; 5g carbohydrate;

2g dietary fiber; 3g net carbohydrate

Directions

1. Simply run your zucchini through the spiral cutter, piling your zoodles into a mixing bowl.

2. Salt the zoodles, tossing as you go. Let them sit for 15 to 20 minutes.

3. Now, use clean hands to squeeze your zoodles to get out the excess liquid and drain well. The liquid will take much of the salt with it.

4. Put a big, heavy pan over medium-high heat and add the olive oil.

5. When it's hot, add the zoodles, tossing just until they are thoroughly good and hot. Stir in the garlic and oregano, and cook for just another minute. Don't let them get mushy.

8. Mac And Cheese

Ingredients

- ½ of a head of cauliflower

- 3 eggs

- 225g cheddar cheese, half tasty, half extra tasty, shredded

- 235ml pouring cream

- 1 teaspoon dry mustard

- ½ teaspoon salt

- ¼ teaspoon pepper

---

6 SERVINGS – Per serving:

323 calories; 29g fat; 13g protein; 2g carbohydrate;

trace dietary fiber; 2g net carbohydrate

Directions

1. Preheat the oven to 180 °C. Grease a large casserole dish.

2. Trim the leaves and the very bottom of the steam off your cauliflower. Now cut the rest into bits about 1.3cm or so.

3. Mix the two kinds of shredded cheddar together. Set aside 60g and layer the rest in the casserole dish starting with the cauliflower.

4. Whisk together the cream, eggs, dry mustard, salt, and pepper and pour it over the cauliflower and cheese. Sprinkle the reserved 60g cheese over the top.

5. Bake for 45 to 50 minutes or until the cauliflower is tender.

9. Broccoli With Anchovy Butter, Olives, And Parmesan

Ingredients

• 340g broccoli

• 55g butter

• 3 anchovy fillets

• 1 clove of garlic, crushed

• 6g chopped blac olives

• 15g shredded parmesan cheese

---

3 SERVINGS – Per serving:

189 calories; 18g fat;

5g protein; 4g carbohydrate;

2g dietary fiber; 2g net carbohydrate

Directions

1. Cut your broccoli into florets. Peel the steams and include them, too. Steam lightly; about 5 to 6 minutes in the microwave is about right.

2. In the meanwhile, put a small saucepan over very low heat and melt the butter. Mince the anchovies and throw them in, along with the garlic.

3. Cook them together, stirring and mashing the bits of anchovy, for about 5 minutes, keeping the heat very low-you don't want to brown the butter or the garlic.

4. When your broccoli is brilliantly green, drain it well and transfer it to a mixing bowl.

5. Pour on the anchovy butter and add the chopped olives. Toss to coat and then plate.

6. Top each serving with 5g of parmesan.

10. Gorgonzola Mushrooms

Ingredients

• 30g butter

• 225g sliced mushrooms

• 1 clove of garlic, crushed

• 50g whipped cream cheese

• 30g Gorgonzola cheese

• Salt and pepper to taste

———————

3 SERVINGS – Per serving:

175 calories; 16g fat; 5g protein; 5g carbohydrate;

1g dietary fiber; 4g net carbohydrate

Directions

1. Put your large, heavy pan over medium heat and melt the butter.

2. Saute the mushrooms, stirring often, until they've softened and turned dark, about 7 to 8 minutes. Stir in the garlic and let it cook for another minute or two.

3. Add the cream cheese and stir until it's melted in. Now, add the Gorgonzola cheese and stir, but only for a minute.

4. Season with salt and pepper, and it's done.

11. Turnips Au Gratin

Ingredients

- 235ml pouring cream

- 55g cream cheese, softened

- ½ teaspoon salt

- ¼ teaspoon pepper

- ¼ teaspoon ground nutmeg

- 3 medium turnips

- 1 small onion

- 225g shredded Colby cheese

---

6 SERVINGS – Per serving:

194 calories; 18g fat;

2g protein; 7g carbohydrate;

1g dietary fiber; 6g net carbohydrate

Directions

1. Preheat the oven to 180 °C. Coat a 2L casserole dish with cooking spray.

2. In a medium-size, heavy-bottomed saucepan, combine the pouring cream, cream cheese, salt, pepper, and nutmeg. Put it over the lowest heat and let it sit while you do the next couple of steps.

3. Trim and peel your turnips and onion and then slice them-thin-I used my food processor's 2mm blade, but you can use a mandolin or a good knife and cutting board.

4. Layer the turnips and onion in the prepared casserole dish-a few layers of turnip, a light layer of onion, then repeat.

5. When your casserole dish is half filled, add a layer of half of the shredded cheese. Continue the layers of turnips and onion, finishing with turnips.

6. Now, go back to that saucepan. Use your stick blender to blend the cream and cream cheese into a sauce. Pour the sauce evenly over the turnips.

7. Top with the rest of the cheese. Cover the casserole dish-with a lid if it has one, or foil if it doesn't-and bake for 40 minutes.

8. Then uncover and bake for another 20 minutes or until bubbly.

12. Loaded Cauli Tots

Ingredients

- ½ a head of cauliflower

- 30g cream cheese, at room temperature

- 15g butter

- 1 teaspoon salt

- ½ teaspoon pepper

- ¼ teaspoon granulated garlic

- 1 egg

- 75g shredded cheddar cheese

- 30g bacon bits

- 1 spring onion, minced

- ½ recipe Pork Rind Crumbs

- Refined coconut oil or MTC oil, for frying

---

4 SERVINGS – Per serving:

257 calories; 18g fat;

18g protein; 6g carbohydrate;

3g dietary fiber; 3g net carbohydrate

Directions

1. Trim the very bottom of the stem and the leaves from the cauliflower and then whack the rest into chunks.

2. Put them in a micro able casserole dish or microwave steamer, add a few tablespoons (45ml) of water, cover, and nuke on high for 10 to 12 minutes or until tender.

3. Drain the cauliflower quite well and put in a deep, narrow bowl.

4. Add the cream cheese, butter, salt, pepper, and granulate garlic and use a stick blender to mash it into Fauxtatoes. Blending the egg.

5. Add the shredded cheese, bacon bits, and spring onion and blend them in, stopping before they've vanished into the mix-you want to be able to see actual bacon bits, shred of cheese, and flecks of spring onion. Now, blend in the Pork Rind Crumbs quickly-the mixture will thicken as you do.

6. In a large, heavy pan, over medium-high heat, melt 1.3cm of coconut oil. You want to bring it to 190 °C; I use a countertop electronic induction burner for this because it lets me hold a precise heat.

7. Drop in the cauliflower mixture in spoonfuls. Fry until a fairly deep brown, flip, and fry the other side. Serve hot.

Chapter 18

SAUCES, CONDIMENTS & DRESSING

1. MAYONNAISE IN A Jar

Ingredients

- 2 egg yolks

- 1 ½ teaspoons lemon juice

- 1 ½ teaspoons wine vinegar

- ½ teaspoon dry mustard

- 2 dashes of Tabasco sauce

- ¼ teaspoon of salt

- 1 to 2 drops of liquid stevia

- 235ml MCT oil

240G

8 servings – Per 30g serving:

257 calories; 29g fat;

1g protein; trace carbohydrate;

trace dietary fiber; negligible net carbohydrate

Directions

I'm putting this first because I devoutly hope you will, indeed, start making your own mayonnaise, especially if you use it often. Because of the MCT oil, this is highly ketogenic and way healthier than the grocery store stuff loaded with soybean oil. It's a snap to make, too. It's far faster than running to the grocery store.

1. First, find a jar the mouth of which will fit your stick blender, such as an old salsa jar.

2. Have your MCT oil measured and standing by in a measuring cup with a pouring lip.

3. Put everything but the oil in the jar. Now, take your stick blender and insert it all the way down to the bottom of the jar. Turn it on and give it a few seconds to blend the egg yolks with the seasonings.

4. Keep the blender running. Now, slowly start pouring in the oil; you want a stream about the diameter of a pencil lead. Work the blender up and down in the jar as you go.

5. When you can't get any more oil to incorporate, and it's pudding on the surface, stop. You're done. Any leftover oil can go back in the bottle. Cap your jar of mayo and stash it in the fridge.

2. Mayonnaise By The Liter

Ingredients

- 2 eggs

- 3 egg yolks

- 15ml lemon juice

- 15ml red wine vinegar

- 1 teaspoon dry mustard

- 10 to 15g Dijon mustard

- 2 dashes of Tabasco sauce

- ½ teaspoon of salt

- 1 teaspoon water

- 700ml MCT oil

———————————

950ML

32 servings – Per 30g serving:

191 calories; 21g fat;

1g protein; trace carbohydrate;

trace dietary fiber; negligible net carbohydrate

———————————

DIRECTIONS

I can use a lot of mayonnaise, especially in warm weather. This has become my go-to mayonnaise recipe-it fills a standard 1 liter mayonnaise jar. You'll need a big food processor-mine holds 3.3l. Why not make it in the jar, as I do with lesser quantities? Because my stick blender won't fit.

1. With the S-blade in place in your food processor, add the eggs, egg yolks, red wine vinegar, lemon juice, mustard, salt, and Tabasco. Turn the processor on. While it's running, add the water.

2. The MCT oil into a glass measuring cup with a pouring lip. With the processor running, add the oil in a thin stream, about the diameter of a pencil lead.

3. When it's all worked in, you're done. Transfer to a tightly lidded jar and stash in the fridge.

4. Theoretically, the shelf life of this is about a week in the fridge, but I've used mine after 10 to 12 days with no ill effects. Your risks are your own to take.

3. Aioli

Ingredients

- 1 egg

- 1 egg yolk

- 30ml lemon juice

- 2 cloves of garlic, crushed

- ¼ teaspoon salt

- 235ml MCT oil

- 120ml olive oil

475ML

16 servings – Per serving:

189 calories; 21g fat; 1g protein; trace carbohydrate; trace dietary fiber; negligible net carbohydrate

Directions

1. With the S-blade in place in your food processor, add the egg, egg yolk, lemon juice, garlic, and salt. Run the processor to combine.

2. With the processor running, add the oils slowly, in a stream about the diameter of a pencil lead.

3. When it's all worked in, you're done. Scrape it into an old 1l mayonnaise jar and stash it in the refrigerator.

4. Tartar Sauce

Ingredients

- 240ml mayonnaise

- 1 sugar free pickle spear, plus 1 teaspoon pickle juice

- 1 teaspoon Dijon mustard

- 1 tablespoon diced red onion

- 30ml lemon juice

- Salt and pepper to taste

300ML

6 servings – Per serving:

266 calories; 31g fat;

1g protein; 1g carbohydrate;

Trace dietary fiber; 1g net carbohydrate

Directions

Just stir everything together. It's nice to make this first and refrigerate it for a while to let the flavor blend.

5. Anchovy Shallot Butter

Ingredients

- 140g butter, at room temperature

- 40g minced shallot (about 1)

- 2 anchovy fillets

- 4g minced parsley

———————————

10 SERVINGS – Per serving:

106 calories; 12g fat; trace protein; 1g carbohydrate; trace dietary fiber; 1g net carbohydrate

Directions

1. In a medium-size pan over medium-low heat, melt 30g of the butter and sauté the shallot until soft and golden.

2. Transfer the shallot and melted butter to the food processor and add the remaining 110g of butter and the anchovy fillets. Process until the anchovies disappear. Add the parsley and pulse until it's mixed in.

3. If you want to be classical about this, turn the mixture out onto a piece of cling film and form into a roll 4cm in diameter. Wrap it up and chill. Then slice nice, pretty round pats to melt over steaks.

4. If that sounds like too much trouble, just keep it in a snap-top container and scoop it out by the spoonful. It's going to melt anyway, you know.

6. Hoisin Sauce

Ingredients

- 60ml MCT oil

- 4 cloves of garlic, crushed

- 160ml soy sauce

- ¼ teaspoon liquid stevia

- 60ml distilled vinegar

- 60g tahini

- 20g chilli garlic sauce

355ML

12 servings – Per serving:

81 calories; 7g fat;

2g protein; 3g carbohydrate;

1g dietary fiber; 2g net carbohydrate

Directions

1. Put a saucepan over medium-low heat, heat the MCT oil, and add the garlic.

2. Saute until lightly golden, about 3 to 5 minutes.

3. Add everything else and stir, working in the tahini. Cook, stirring often, for about 5 minutes.

4. Store in a tightly lidded jar in the refrigerator.

7. Bearnaise Sauce

Ingredients

- 60ml dry white wine

- 2 egg yolks

- 15ml lemon juice

- 60ml white wine vinegar

- 4g minced fresh tarragon

- 10g minced shallot

- 115g butter

---

4 SERVINGS – Per 30g serving:

247 calories; 26g fat;

2g protein; 2g carbohydrate;

0g dietary fiber; 2g net carbohydrate

Directions

1. In a small saucepan over low heat, reduce the wine to about 15ml.

2. Put it in the top of a double boiler, along with the vinegar, fresh tarragon, minced shallot, and egg yolks. Have your butter measured-I think it's easier if you melt it.

3. Okay, it's show time: Put the double boiler over simmering water.

4. Turn on your stick blender, immerse it in the egg yolks and seasoning, and blend them up.

5. Keep blending. Start adding the butter about 20g at a time, blending the whole tine. Keep blending for about 30 seconds between each addition of butter. Your sauce should start thickening and fluffing up. Keep blending.

6. When all the butter is in, keep blending for another minute or two, making sure your sauce is good and thick before you turn off the stick blender and take the pan off the heat. If it's not finished cooking, your sauce will fall apart.

7. Once it's done, it will be lovely over steak, and any leftovers will keep for a day or two in a snap-top container in the fridge.

8. Porcini, Portobello, And Button Mushrooms In Cream

Ingredients

- 15g dried porcini mushrooms

- 120ml boiling water

- 30g butter

- 30ml extra virgin olive oil

- ½ of a medium onion, diced

- 115g Portobello mushrooms, sliced

- 115g button mushrooms, sliced

- 5 cloves of garlic, crushed

- 2 teaspoons minced fresh rosemary

- 8g minced parsley

- 60ml dry red wine

- 60ml cream

-]½ teaspoon salt

- ¼ teaspoon pepper

475ML

8 servings – Per serving:

104 calories; 9g fat;

1g protein; 4g carbohydrate;

1g dietary fiber; 3g net carbohydrate

Directions

1. First, put your porcini in a dish and pour the boiling water over them. Let them sit for 10 minutes or so.

2. In the meanwhile, put your large, heavy pan over medium-low heat and add the butter and olive oil, swirling them together as the butter melts.

3. Add the diced onion, then the Portobello and button mushrooms. I'm assuming you bought them sliced. Use the edge of your spatula to break them up a bit more as they sauté.

4. Use a fork to lift the soaked porcini out of their water-reserve the water-chop them medium-coarse and add them to the other mushrooms.

5. As the fresh mushrooms start to soften, add the garlic, rosemary, and parsley. Keep sautéing, stirring often, until the mushrooms are dark and soft.

6. Now, add the wine and the soaking water from the porcini. Stir everything up, turn up the heat to medium, and let the mixture boil until most of the liquid has boiled away. Then, turn the heat back down.

7. Stir in the cream, salt, and pepper. Let the whole thing cook for another minute or two. Done.

9. White Sauce

Ingredients

- 355ml cream

- 30g butter

- 115g cream cheese, cut into small cubes

- Salt and pepper to taste

---

6 SERVINGS – Per serving:

305 calories; 32g fat; 3g protein; 2g carbohydrate; 0g dietary fiber; 2g net carbohydrate

Directions

This is the keto version of the béchamel, the classic white sauce. If you need it thicker than this, you can add a sprinkle of xanthan gum.

1. Simply combine everything in a heavy-bottomed saucepan over low heat. Whisk as it heats until the cream cheese has melted in entirely.

10. Sauce Mornay

Ingredients

- 235ml cream

- 10g minced shallot

- 1 egg yolk

- 15g shredded Gruyere cheese

- 10g shredded Parmesan cheese

- 1 dash Tabasco sauce

- ¼ teaspoon salt

285ML

5 servings – Per serving:

197 calories; 20g fat;

3g protein; 2g carbohydrate;

trace dietary fiber; 2g net carbohydrate

Directions

This de-carbed version of a classic sauce is beyond luscious. I'ts great on asparagus, broccoli, eggs, fish-you name it.

1. In a double boiler over hot but not boiling water, start the cream warming and add the shallot.

2. When the cream is just below simmering, whisk 60ml of it into the egg yolk. Then, whisk the cream and yolk mixture back into the cream. Do not try to simply whisk the yolk into the main pot of cream.

3. Whisk in both chesses and keep whisking as the sauce thickens, which will happen quite quickly. Stir in the hot sauce and salt, and your sauce is done. Try to refrain from simply eating it all out of the pot.

11. Mushroom Bordelaise

Ingredients

- 40g butter, divided

- 1 shallot, minced

- 10g diced onion

- 1 clove of garlic, crushed

- 235ml beef stock

- 80ml dry red wine

- 8 peppercorns

- 1 whole clove

- 1 bay leaf

- ¼ teaspoon fresh thyme leaves

- 225g sliced mushrooms

- 15 ml Worcestershire sauce

- Xanthan gum

- Salt and pepper

700ML

8 servings – Per serving:

61calories; 5g fat;

1g protein; 3g carbohydrate;

1g dietary fiber; 2g net carbohydrate

Directions

1. In a small saucepan over medium-low heat, melt 15g of the butter. Add the shallot and onion and sauté until softened and start to brown, about 5 to 7 minutes. Add the garlic and sauté for another minute or two.

2. Add the beef stock, wine, peppercorns, clove, bay leaf, and thyme. Turn the heat up to medium. Bring to a simmer and turn the heat down to hold it there. Let cook for 7 to 8 minutes.

3. While that's happening, melt the remaining butter in a large saucepan over medium heat. Add the mushrooms and sauté, breaking them up further with your spatula, until they've softened, changed color, and exuded their liquid. Conveniently, this should take roughly as long as the stock/wine mixture needs to simmer.

4. When the mushrooms are sautéed and the stock/wine mixture has simmered, put a strainer over the saucepan with the mushrooms in it and strain the liquid into it. Stir in the Worcestershire. Thicken to the texture of pouring cream with your xanthan gum and season with salt and pepper to taste.

12. Maitre D'Hotel Butter

Ingredients

- 110g butter, at room temperature

- 2 teaspoons lemon juice

- 4g minced parsley

---

8 SERVINGS – Per serving:

102 calories; 11g fat; trace protein; trace carbohydrate; trace dietary fiber; negligible net carbohydrate

Directions

This is a classic and one of the secrets of top-notch steakhouses. Throw this on a good grilled rib eye or sirloin, and what more could you want? Maybe a glass of dry red?

1. Put the butter in your food processor and run until it's creamy. Add the lemon juice and run until it's evenly blended in. Add the parsley and pulse just until it's mixed in; you don't want to pulverize it.

2. Melt a good dollop of this on each steak as it comes off the heat.

3. If you make this ahead of time, you can scrape it out onto a piece of cling film or wax paper, form it into a tidy roll, and stash it in the fridge.

4. Then, you can slice pretty pats of it to melt on your steaks. It's not only tasty, but impressive, should you have anyone you'd like to impress.

13. Beurre Noir

Ingredients

- 235ml cream

- 10g minced shallot

- 1 egg yolk

- 15g shredded Gruyere cheese

- 10g shredded Parmesan cheese

- 1 dash Tabasco sauce

- ¼ teaspoon salt

This ultra-simple sauce is actually classical French cooking. It's a good way to up the fat content-and flavor of fish, but try on fried eggs, too.

---

8 SERVINGS – Per serving:

102 calories; 11g fat;

trace protein; trace carbohydrate;

trace dietary fiber; 0g net carbohydrate

Directions

1. In a saucepan over medium-low heat, melt the butter and cook it until it browns. Stir in the lemon juice, season with salt and pepper to taste, and you're done.

14. Remoulade

Ingredients

- ½ of a shallot

- 17g capers, drained

- 225g mayonnaise

- 1 teaspoon lemon juice

- 4g chopped parsley

- 4g chopped fresh tarragon

- 9g chopped dill

- 2 anchovy fillets

---

8 SERVINGS – Per serving:

201 calories; 23g fat; 1g protein; trace carbohydrate; trace dietary fiber; negligible net carbohydrate

Directions

A classic of French cuisine, remoulade has become a world traveler. Most commonly, it is used on fish, but depending on the region, it is also served with beef, French fries, celery root, and even hot dogs. Surely, you can find a use.

1. Put the shallot, capers, parsley, tarragon, dill, and anchovies in your food processor and pulse until it's all quite finely chopped.

2. Add the lemon juice and mayonnaise and run just long enough to blend.

3. Refrigerate for an hour or two before serving to let the flavors marry.

15. Wasabi Mayonnaise

Ingredients

- 115g mayonnaise

- 2 teaspoons soy sauce

- 3 drops of liquid stevia

- 1 teaspoon lemon juice

- 1 teaspoon wasabi paste

120ML

8 servings – Per serving:

100 calories; 12g fat;

trace protein; trace carbohydrate;

trace dietary fiber; negligible net carbohydrate

Directions

Just combine everything in a bowl and whisk together well. It's unbelievably good.

16. Caesar Dressing

Ingredients

- 2 to 4 anchovy fillets, to taste

- 30g Dijon mustard

- 15ml white wine vinegar

- 30ml Worchestershire sauce

- 2 cloves of garlic, crushed

- 15ml lemon juice

- 1 egg

- 120ml extra-virgin olive oil

- 25g grated parmesan cheese

235ML

6 servings – Per serving:

198 calories; 20g fat;

3g protein; 2g carbohydrate;

Trace dietary fiber; 2g net carbohydrate

Directions

1. Assemble everything from the anchovies through the egg in your food processor, with the S-blade in place. Run for a minute or so until the anchovies are pulverized.

2. Now, with the processor running, slowly pour in the olive oil in a stream about the diameter of a pencil lead. Your dressing should emulsify.

3. Add the parmesan and pulse to mix it in. That's it. Store in a tightly lidded container in the fridge and use it up within a few days because of that egg.

17. Ranch Dressing

Ingredients

- 120ml buttermilk

- 115g sour cream

- 60g mayonnaise

- 25g spring onion

- 1 or 2 cloves of garlic, crushed

- 2 teaspoons lemon juice

- 1 ½ teaspoon dried dill

- ½ teaspoon salt

- ¼ teaspoon pepper

- 2 dash of Tabasco sauce

———————

MAKE 355ML

12 servings – Per serving:

63 calories; 6g fat; 1g protein;

1g carbohydrate; trace dietary fiber; 1g net carbohydrate

———————

DIRECTIONS

Ranch dressing is beloved by many, but bottled varieties too often have bad oils. We won't even talk about the fat-free stuff, which is pretty much spicy corn syrup. This is easy and will get some veggies into your kids.

1. It's so simple. Assemble everything in your blender and process until it's well blended. Chill for a least a few hours for the flavors to combine.

18. Classic Coleslaw Dressing

Ingredients

- 115g mayonnaise

- 115g sour cream

- 1 teaspoon wholegrain mustard

- 15ml apple cider vinegar

- 8 drops of liquid stevia

- ½ teaspoon salt or to taste

———

MAKE 235ML

5 servings – Per serving:

208 calories; 24g fat;

1g protein; 1g carbohydrate;

Trace dietary fiber; 1g net carbohydrate

Directions

1. Measure everything and stir it all together in a bowl. Done. Once you've done this a few times, you'll be able to just eyeball it.

———

19. BUFFALO Slaw

Ingredients

- 350g shredded cabbage (about ¼ of a head)

- 5 spring onions, sliced

- 120g thinly sliced celery

- 115g mayonnaise

- 60ml of Tabasco sauce

- 1 clove of garlic, crushed

- 115g crumbled mild blue cheese

---

6 SERVINGS – Per serving:

222 calories; 21g fat;

6g protein; 6g carbohydrate;

2g dietary fiber; 4g net carbohydrate

Directions

1. Combine the cabbage, spring onions, and celery in a mixing bowl.

2. In a small bowl, stir together the mayonnaise, Tabasco sauce, and garlic.

3. Pour over the vegetables and stir until they're evenly coated.

4. Add the blue cheese and gently stir it in-you want there to be actual chunks of blue cheese in your slaw.

You can serve this right away, but consider chilling for an hour or two to let the flavors blend.

Chapter 19

DESSERTS

1. PUMPKIN CHEESECAKE With Ginger-Pecan Crust

Ingredients

CRUST

- 200g pecan halves

- 50g erythritol

- 1 ½ teaspoons ground ginger

- 55g butter, chilled

- 15ml water

- 12 drops of liquid stevia

FILLING

- 245g canned pumpkin

- 120ml cream

- 2 teaspoons vanilla extract

- 1 teaspoon ground cinnamon

- ½ teaspoon liquid stevia

- ½ teaspoon ground ginger

- ¼ teaspoon ground nutmeg

- ¼ teaspoon salt

- 680g cream cheese, softened

- 95g powdered erythritol

- 1 teaspoon molasses

- 4 eggs

Dulce de leche

Whipped Cream, for serving (below)

---

12 SERVINGS – Per serving:

420 calories; 41g fat;

8g protein; 8g carbohydrate;

2g dietary fiber; 6g net carbohydrate

Directions

1. To make the crust, preheat the oven to 180°C, positioning the rack in the middle. Coat a 23cm springform pan thoroughly with cooking spray.

2. Put the pecan halves, erythritol, and ground ginger in your food processor and run until the pecans are ground to a meal.

3. Cut the butter into 4 tablespons (14g), add, and pulse until it's cut in. Mix the water and liquid stevia in a cup. With the processor running, drizzle the mixture in.

4. When a dough forms, turn off the processor. Turn the dough out into the prepared pan and press it firmly and evenly across the bottom, making sure to cover the seam at bottom edges.

5. Bake the crust for 12 to 14 minutes. Let it cool while you make your filling.

6. To make the filling, in a medium-size mixing bowl, whisk together the pumpkin, cream, vanilla, cinnamon, liquid stevia, ginger, nutmeg, and salt.

7. In a large mixing bowl, use your electric mixer to beat the cream cheese until quite creamy, about 2 to 3 minutes.

8. Scrape down the sides of the bowl often. Beat in the erythritol and molasses, blending very well.

9. Add the pumpkin mixture and mix until well blended. Then, beat in the eggs, one at a time, still scraping down the sides of the bowl often to make sure everything is evenly blended.

10. Tear off a big piece of aluminium foil-about 50cm long. Wrap the outside of your springform pan with it-you'll be putting it in a water bath, and you don't want water leaking in through the seams.

11. Scrape the batter into the foil-wrapped pan. Place it in a larger roasting pan and pour hot water around the springform to about 2,5cm deep. Place the whole shebang in the oven.

12. Bake for 60 to 70 minutes or until mostly set but still a little jiggly in the center. Carefully, lift the springform from the water bath and place it on a wire rack to cool.

13. Grab a thin-bladed paring knife. Run it around the edge of the cake, loosening it from the sides. This should prevent cracking as it cools. When cool, chill well. Serve with Dulce de Leche Whipped Cream.

---

2. DULCE DE LECHE WHIPPED Cream

Ingredients

• 30ml sugar free coffee flavoring syrup, caramel

• 235ml cream

---

12 SERVINGS – Per serving:

68 calories; 7g fat; trace protein; 1g carbohydrate; 0g dietary fiber; 1g net carbohydrate

Directions

1. Put your caramel syrup in a small saucepan and cook over low heat until reduced to 1 tablespoon (15ml). Mix with the cream and then chill the combination for several hours.

2. Whip as you would regular whipped cream.

3. Flan

Ingredients

- 475ml cream

- 65g erythritol

- 6 eggs

- ½ teaspoon liquid stevia

- 1 teaspoon vanilla

- 1 pinch of salt

- 1 pinch of ground nutmeg

- 90ml sugar-free caramel coffee flavoring syrup

---

6 SERVINGS – Per serving:

341 calories; 34g fat;

7g protein; 3g carbohydrate;

trace dietary fiber; 3g net carbohydrate

Directions

This would make a great summer breakfast.

1. Preheat the oven to 180°C. Grease a 25cm pie plate.

2. Put the cream, eggs, erythritol, liquid stevia, vanilla, salt, and nutmeg in your blender and run until all well combined.

3. Put a shallow baking pan on the oven rack. Place the prepared pie plate in it. Pour water into the outer pan up to within about 1.3cm of the rim of the pie plate.

4. Now, pour the custard mixture into the pie plate. Bake for about 50 to 60 minutes before chilling.

5. You can run a knife around the edge and invert the flan onto a plate and then top with the caramel syrup to serve, but it's easier to just cut wedges like a pie or spoon it out. Still, top it with the syrup to serve.

4. Keto Brownies

Ingredients

- 55g bitter chocolate

- 225g butter

- 95g erythritol

- ½ teaspoon liquid stevia

- ½ teaspoon vanilla extract

- 2 eggs

- 115g vanilla whey protein powder

- 1 pinch of salt

———————

12 SERVINGS – Per serving:

208 calories; 19g fat; 9g protein; 2g carbohydrate; 1g dietary fiber; 1g net carbohydrate

Directions

1. Preheat the oven to 180°C. Coat a 20x 20cm baking dish with cooking spray.

2. In the top of a double boiler over boiling water or in a saucepan over a heat diffuser set over very low heat, melt the chocolate and the butter together. Stir until they're well combined. Scrape this into a mixing bowl.

3. Add the erythritol, stir well, and then stir in the liquid stevia, and vanilla extract. Next, beat in the eggs, one at time. Stir in the vanilla whey protein powder and salt.

4. Pour into the prepared baking pan. Bake for 15 to 20 minutes. Do not overbake. Let cool in the pan and then cut into 12 bars. Store in a tightly covered container in the refrigerator.

5. Coffee Mousse

Ingredients

- 1 ½ teaspoons gelatine powder

- 50g erythritol

- 30ml cold water

- 45ml boiling water

- 235ml strong brewed coffee

- ¼ teaspoon liquid stevia

- 1 tiny pinch of salt

- 385ml of coconut milk, chilled

---

4 SERVINGS – Per serving:

191 calories; 20g fat;

3g protein; 3g carbohydrate;

0g dietary fiber; 3g net carbohydrate

Directions

1. Put the gelatine in a small dish and add the cold water. Let it sit for a few minutes until all the water has absorbed and the gelatine dissolves.

2. In a medium-size mixing bowl, combine the gelatine with the coffee, erythritol, liquid stevia, and salt.

3. Wisk until the erythritol is dissolved and everything is well combined. Put the bowl in the refrigerator-you want to chill it until it's the texture of egg white.

4. When the gelatine mixture has thickened, you're ready to continue. Put your coconut milk into a small, deep mixing bowl and whip it on high speed until it's fluffy and thickened.

5. Don't expect it to turn out like whipped cream; it won't get that stiff. But do whip it for a good 4 to 5 minutes on high.

6. Grab your bowl of gelatine and use the same beaters to whip it until it, too, is fluffy and thickening.

7. Now, with the mixer running, add the coconut milk to the gelatine mixture. Whip them together. Spoon the mixture into 4 dessert dishes and chill for a least several hours, and overnight is great.

6. Courgette Cake

Ingredients

- 170g hazelnuts

- 115g vanilla whey protein powder

- 60g Virtue sweetener (mix erythritol and monk fruit)

- 1 ½ teaspoons ground cinnamon

- 1 teaspoon baking soda

- ½ teaspoon salt

- ¼ teaspoon ground nutmeg

- 77g full-fat Greek yogurt

- 28ml pourable coconut milk

- 2 eggs

- 120ml MCT oil

- 120g shredded zucchini

---

8 SERVINGS – Per serving:

345 calories; 30g fat;

16g protein; 6g carbohydrate;

2g dietary fiber; 4g net carbohydrate

Directions

1. Preheat your oven to 170°C. Coat a 23cm pan with cooking spray.

2. Put the hazelnuts in your food processor and run until they have the texture of cornmeal. Transfer to a mixing bowl.

3. Use 15g of the hazelnut meal to 'flour' your cake pan. Add the vanilla whey protein, virtue, cinnamon, baking soda, salt, and nutmeg to your food processor.

4. Pulse until everything is well combined. Dump the mixture into the mixing bowl with the ground hazelnuts.

5. In a separate bowl, combine your yogurt, coconut milk, eggs, and MCT oil. Whisk these together well.

6. Pour the liquid ingredients into the dry ingredients and whisk until you're sure you have no lumps or pockets of dry stuff left. Stir in the zucchini. Pour the batter into the prepared pan.

7. Bake for 70 minutes until a tester inserted near the center comes out clean. Let cool in the pan and then turn out onto a wire rack to cool completely.

7. Budino Di Cioccolato

Ingredients

- 475ml cream

- 175g sugar-free chocolate chips

- 2 egg yolks

- 30ml dark rum

---

12 SERVINGS – Per serving:

208 calories; 19g fat; 9g protein; 2g carbohydrate; 1g dietary fiber; 1g net carbohydrate

Directions

1. Put the cream in a heavy-bottomed saucepan over low heat and bring just to a simmer.

2. In the meanwhile, combine the chocolate chips, egg yolks, and rum in your blender.

3. When the cream is hot, start pouring it into the blender, turning on the blender when it's about half in. Run the blender until the chocolate is all melted.

4. Pour it into 6 little dishes, because it's super-rich.

5. Chill for several hours, preferably overnight.

www.ingramcontent.com/pod-product-compliance
Lightning Source LLC
Chambersburg PA
CBHW071250220526
45468CB00001B/61

* 9 7 8 1 9 8 3 0 4 7 8 8 6 *